GEOFFREY MARNELL

Mathematical Doodlings

Curiosities, conjectures and challenges

Burdock Books

First published in Australia in 2017 by
Burdock Books
www.burdock.com.au
info@burdock.com.au

National Library of Australia Cataloguing-in-Publication entry

Creator: Marnell, Geoffrey R., author.

Title: Mathematical doodlings : curiosities, conjectures and challenges/ Geoffrey Marnell.

ISBN: 9780994366696 (paperback)

Subjects: Mathematical recreations.

            Mathematics--Miscellanea.

Cover design by Eugeniy Sosunov

To my wife *Melinda Lancashire,* whose
generous dispensations afforded me the
time to write this work.

By the same author

Contents

Appendices 139

$$(n) \sim \frac{\sqrt{2n}}{\log n} \cdot \prod_{v=3} \left(1 - \frac{1}{v-1} \cdot \left(\frac{2n}{v}\right)\right)$$

Introduction

Doodling is not a noun in the English language. But using it as a noun serves my purpose well. It enables me to position the offerings in this book by analogy: doodlings are to doodles what ducklings are to ducks—small, imperfectly formed but with potential to grow into something more. There are drawings—expertly crafted representations— and there are doodles: idle patterns, often sketched absent-mindedly, but not always without some visual interest. To equate the petite mathematical discoveries recorded in this book with drawings would be outrageously pretentious; to equate them with doodles would be presumptuous. Rather, they are of a lesser breed—more graceless swirl than elegant spiral. Which naturally invites the question of why I am putting them into the public arena. There are two reasons. The first is to inspire the timid lover of numbers to embrace the joy of discovery undaunted by the seemingly unfathomable complexities entertained by professional mathematicians. This is an activity so much more likely to prove fruitful today thanks to the advent of cheap computing. You don't need the sledgehammer tools of professional mathematics to spot the odd speck of gold in your prospecting pan. A common-or-garden spreadsheet application, a simple computer-based algebra system and a modicum of patience is all you need. Indeed, the timid numerophile should be heartened by the respect that amateurs have earned in other fields. Take astronomy, for instance. Amateur astronomers have spotted stars that professional astronomers have missed, despite the latter's billion-dollar telescopes. So the amateur mathematician might well spot a hitherto unseen wanderer, black hole or comet in the vast mathematical universe. Its significance might not be immediately obvious, but perhaps a greater mind might be drawn to the discovery and embellish it into, if not a drawing, then at least an interesting doodle. The large is born small; the duck once a duckling. Which brings me to the second reason for publishing this work: the hope—born, perhaps, of a madman's solipsistic excitement—that there might be something, even one thing, significant in what I have idly discerned in the kaleidoscopic richness of the mathematical universe.

There are many factors that might lead the amateur to doubt the worth of their endeavours, thus giving timidity a hand over discovery. For a start, there is the prospect of embarrassment on finding that what is claimed as one's own discovery has already been footnoted in the annals of mathematics. Whatever the discipline, it is understandable that the amateur will harbour this worry. It's a worry that dampens, if not crushes, their enthusiasm. There are two counterpoints we can make here. Firstly, there is joy to be had in producing an artful drawing of, say, a cat even though cats have been drawn before (and for thousands of years). There is even pleasure to be had in seeing art and hearing music that is in some way derivative (that is, of a style that is not original). I can enjoy the paintings of Braque despite seeing the unmistakable influence of Picasso. I can enjoy the music of Prokofiev despite hearing echoes of Stravinsky. In other words, one can admire the *technique* of creation—its execution—without needing to be impressed by its *originality*. Thus the joy of a mathematical discovery—and the pride elicited by stumbling on a novel method of discovery—need not be diminished by the possibility that others have beaten us to the discovery. Secondly, there is many a precedent: some of what professional mathematicians have discovered had already been discovered. To take one example: in 1963, Scottish mathematician Hyman Levy published a conjecture known for a while as *Levy's conjecture* (namely that all odd integers greater than 5 can be expressed as the sum of an odd prime number and an even semiprime[1]). Levy was clearly unaware that the French mathematician Émile Lemoine had made the same conjecture earlier, in 1895, and, in recognition of that precedence, the conjecture is now better known as *Lemoine's conjecture*. If the professional mathematician can propose what turns out not to be original, surely the amateur mathematician should be granted the same licence.

Nor should the would-be doodler be dissuaded by the Zeitgeist of today, with its unhealthy emphasis on all things mercantile. Modern *Homo sapiens* has, it seems, created a culture for itself that values only what can be measured in monetary terms. If it is quick, as it has been, to devalue the *literae humaniores*, it is sure to be even quicker in declaring number-play a useless activity. Economists—today's high priests—would consider it an irrational activity, for it displaces the utility-seeking that is stamped on our genes by evolution (and thus, supposedly, instinctual). Sadly, this corrosive culture has began to permeate universities, once powerhouses of research for the sake of it, where wastepaper baskets brimmed with discarded doodlings and doodles. Today university research is increasingly industry-directed (and

---

1. A semiprime is the product of two prime numbers.

thus profit-driven). It has become more about exploiting
what we already know than exploring what we don't
know. And yet exploring what we don't know is what has
delivered us from the nasty, brutish and short life of the
Neanderthal. And pure mathematics—unadulterated by
profit margins, returns on investment and shareholder
entitlement—has been the riverhead of much that we now
enjoy. Without the idle, number-playing of Newton and
Leibnitz, we might never had stumbled upon the
mechanisms of calculus that underpin much of modern
life. Without the idle exploration of prime numbers, we
might never have developed the robust techniques of
encryption that enable secure transmission of electronic
data. Without ... the list goes on. So you might never
know what becomes of your idle doodling: the new
calculus of tomorrow; the new encryption of the century
before us. And if nothing significant seems likely to come
of your endeavours, you can still challenge an accusation
of irrational time-wasting on the grounds that mathematical doodling has *indirect* utility. As the British
philosopher Bertrand Russell wrote:

> "there is indirect utility, of various different kinds, in the
> possession of knowledge which does not contribute to
> technical efficiency ... Perhaps the most important
> advantage of 'useless' knowledge is that it promotes a
> contemplative habit of mind ... [and] a contemplative
> habit of mind has advantages ranging from the most
> trivial to the most profound."[2]

Nor should the doodler be discouraged by criticism of
the typical method of doodle discovery, namely, *inductive
reasoning*. It is true that professional mathematics has
been—and still largely remains—an endeavour of a

2. Bertrand Russell, *In Praise of Idleness and Other Essays*, Routledge,
    London, 1935, pp. 21, 24.

different form of reasoning, namely, *deductive reasoning*. In deductive reasoning, a new claim is *deduced* from one or more axioms or previously established truths. If the premises are true, the conclusion is necessarily true. In deductive reasoning we typically argue from the general to the specific: All men are mortal (general), Socrates is a man, therefore Socrates is mortal (specific). Inductive reasoning is the mirror image of deductive reasoning. Here we argue from the specific—or at least numerous instances of the specific—to the general: drug x cured disease y in patient z_1, drug x cured disease y in patient z_2, drug x cured disease y in patient z_3 (specific instances) ... therefore drug x is a cure for disease y (general statement). Traditionally, deductive reasoning has been the tool of mathematicians and philosophers; inductive reasoning the tool of scientists.

But the advent of computers has brought with it a new sub-discipline of mathematics: *inductive mathematics*[3]. Inductive mathematics follows the scientific method. Practitioners gather lots of instances that support some hypothesis—for example, every whole number greater than 5 can be expressed as $p + (q \times r)$ where p, q and r are prime numbers—and, at some reasonable point, declare the hypothesis to be true. (This is a slight simplification, but no greater detail is needed here to make our point.) And this is the approach taken in this book. There are some deductions, but mostly what I am proposing are conjectures that grew out of examining numerous instances of some pattern or other. And the examination was done mostly by iterative computer analysis (using nothing more sophisticated than common-or-garden database and spreadsheet applications).

---

3. This is not to be confused with *mathematical induction*, which still relies on deduction as its engine.

Now there are those who argue that inductive reasoning can never give us truth. It is, so its detractors claim, the bastard cousin of deductive reasoning. Faced with such a judgement, the timid doodler might be prone to give up. But it is worth untying the threads that give this claim its apparent strength. It is true, in one sense, that inductive reasoning doesn't *prove* anything. But it can justify us in believing that something is true. Inductive knowledge is like an asymptote. Repeated verification (or failed falsification) brings us ever closer to truth even though we can never reach it (just as the graph of $y = x^2/(x - 1)$ gets ever so closer to the line $y = (x - 1)$ without ever touching it). There is always some chance that a falsifying instance will be found. (Patient z_{654} was *not* cured of disease y with drug x.) So inductive reasoning does not establish truths *beyond all doubt*, a fact often seized upon by science deniers and unscrupulous lawyers[4]. But it can still give us *knowledge*—at least as the word is commonly used. If that were disallowed, then *all* of what we now accept as scientific knowledge would not be knowledge. Take, for instance Ohm's Law, which states that the current in an electrical circuit is proportional to the voltage and inversely proportional to the resistance. Every morning, as billions of electrical devices are turned on around the world, Ohm's Law is verified. Maybe there is an extremely rare condition, not yet observed by scientists, where Ohm's Law does not hold, but the billions of verifications made every day surely allow us to claim that we *know* that Ohm's Law is *true*.

---

4. In 2012, seven people, including four scientists, were sentenced to six years imprisonment in Italy for failing to predict a 2009 earthquake in which more than 300 people died. The judge clearly (but erroneously) believed—or was persuaded to believe—that science was a deductive rather than probabilistic endeavour.

And that is so whatever the inductive endeavour. It does violence to our language, tears at the common meaning of *knowledge* and *truth*, to argue that since inductive reasoning does not carry the same immediate, unfalsifiable force as deductive reasoning—because there is always the chance of a falsifying instance—that there can be no inductive knowledge or inductive truth. We speak without contradiction of *knowing* that the sun will rise tomorrow, that the bacterium *Helicobacter pylori* causes stomach ulcers, that unaided, water seeks out points of lower altitude. And yet deductive reasoning played no major part (or any part) in our reaching that knowledge. So we should not be disheartened in our quest for mathematical knowledge using inductive means because an induction-based claim might one day be falsified. Still, inductive reasoning can never be a proof as mathematicians and philosophers understand proof. At best there can be *conjectures*—and this book offers many.

And needless to say, even longstanding and seemingly well-founded conjectures can turn out to be false. One example: in 1752, Prussian mathematician Christian Goldbach conjectured that every odd integer is either a prime number or could be expressed as the sum of a prime number (including 1) and twice the square of an integer (in other words, as $p + 2a^2$). The conjecture looked good: the Swiss mathematician Leonard Euler could not find an exception after checking 2500 numbers. But a little over a century later—after a pencil-and-paper slog through the odd integers up to 9000—German mathematician Moritz Stern announced that he had found exceptions—*just two*: 5777 and 5993. Oddly, no more exceptions have been found even with the benefit of computer analysis.[5]

Conversely, the humble conjecture, when given enough attention, might turn out to be justified by way of a deductive proof. Take *Fermat's Last Theorem*. In the margins of a copy of *Arithmetica*, written by the Ancient Greek mathematician Diophantus, Fermat wrote that he had a proof that no three positive integers p, q and r can satisfy the equation $p^n + q^n = r^n$ if n is greater than two. Many scholars have argued that this really should have been called *Fermat's Last Conjecture* because Fermat did not show how a proof might be possible. This view seems entirely correct. Fermat claimed, in the same marginal note, to have a proof (stating that it was too complex to fit into the margins of the book). Oddly, he never subsequently wrote down his proof, which adds weight to the view that his so-called theorem should have been called a conjecture. Even so, his conjecture has been proved (by Andrew Wiles in 1994, 357 years after Fermat penned his marginalia), which shows that conjectures can be proved just as they can be disproved. (Indeed, this is where inductive mathematics differs from science. The inductive conjectures of science — more commonly known as *hypotheses* — can never be proved deductively; the inductive conjectures of mathematics might well be.) So there is no reason to be disheartened if all you have come up with — or can come up with — is a conjecture (providing, of course, that you have shown that some pattern or other in the mathematical universe appears to be governed by it). It may well turn out, given enough attention, to be a *proven* truth.

---

5. See Laurent Hodges, "A Lesser-Known Goldbach Conjecture", *Mathematics Magazine*, vol. 66, no. 1, 1993, pp. 45–47. Hodges had computer-analysed the first million odd integers and found no further exceptions. He has christened 5777 and 5993 *Stern numbers*.

§

The bookends of this work discuss intelligence tests. My suspicion of these tests began early. At the time I was growing up, many in the medical profession considered stuttering—an affliction I had and still have—to be a symptom of mental retardation. So at the start of the treatment my mother sought for me I was given an intelligence test. I must have passed the test, as I wasn't institutionalised. But I remember sensing a profound dislike of it. Some of the questions on the test—such as what is the capital of Greece—clearly belonged in the sphere of knowledge, not intelligence. (Everyone in Greece—genius or dullard—would have got that question correct, I remember thinking.[6]) But even some of the questions that did require reasoning, as opposed to mere regurgitation of facts, worried me, most notably the numeracy questions. Even with only primary school arithmetic behind me, I felt certain that some of these questions admitted to more than one answer. I recall feeling aggrieved that my seemingly legitimate answers might be marked wrong, unfairly pushing me leftwards along the bell curve, if not into the wretched territory of the idiot, then certainly into the no less unattractive company of the imbecile. (Yes, "idiot" and "imbecile" were the classificatory bins into which the early psychometricians dumped those whose test scores were sub-normal. How deaf could they have been to the life-scarring connotations of those words.) Happily, my fears

---

6. Confusing knowledge with intelligence continues to this day. A question on a contemporary IQ test asks "The acronym RSVP originally came from the French term *Répondez s'il vous plaît* – True or False?" See http://testingcentre.nicheconsulting.co.nz/iq_tests.htm. Viewed 16 December 2016. (By the way, an intelligent—sorry, knowledgeable—person would know that *RSVP* is an initialism, not an acronym.)

of imbecility were ungrounded. But my doubts about the validity of intelligence tests—whether they actually measure what they purport to measure—were to return later, when psychometrics became the new craze in human resource departments. Your skill at persuasively answering questions at a job interview would now be qualified by the score you got on an intelligence test (euphemistically renamed *aptitude test*). It was no longer enough to prove you could do the work required of the position with both hands tied behind your back. You also had to be good at knowing that library is to book what book is to page, that the initialism *RSVP* stands for *répondez s'il vous plaît* and that the next number in the sequence 4, 5, 8, 17, 44 is 125.

The fetish for psycho-quantification also seeped into education. Admission to selective-entry schools and colleges, and the offer of a scholarship, increasingly became dependent on getting a good score on an intelligence test. My childhood grievance that I might easily have been misclassified now morphed into a stronger sense of general unease. It seemed to me that natural justice was being denied on a grand scale. For a start, if intelligence tests were in part tests of *knowledge*, then an assiduous student from a poorly resourced school would have less chance of gaining a coveted scholarship than a lackadaisical student from a well-resourced private school who happened to have been spoon-fed more units of information over the same period of schooling. How could that be fair? And if, as I had strongly suspected, some designers of intelligence tests lacked the cognitive wherewithal to devise questions that admitted to only one answer, it seemed quite possible that bright subjects, in giving legitimate answers that were marked as incorrect, were being unfairly down-graded.

The confluence of my love of numbers and suspicion of intelligence tests led me to idle away many hours seeking alternative answers to the numeracy questions I found on such tests. It was a fun way to spend a wet Sunday, and it gave birth to a life-long hobby of number doodling. Perhaps the most common type of numeracy question one finds on IQ tests is the number-sequence problem. An example is {4, 5, 8, 17, 44 ?}. I had a lot of fun with these, discerning patterns and relationships that clearly the test designer had not seen. Indeed, a little doodling soon convinced me that if all that the test-setter was looking for was some rule that could generate all the given numbers in the given order, then the only correct answer to any number-sequence question was "any number". For no matter how the sequence is continued — with 125, 8531, 75 or whatever — there is an easily discoverable rule that would generate the sequence as it was continued. Indeed, as I show in the first part of this book, an *infinite* number of rules can generate it. At last I could give grounds to my suspicion that number-sequence questions are poor discerners of intelligence and should be banished from intelligence tests. The mad excitement that bubbled up in me at the time of this discovery spurred me on for years to explore as much of the mathematical universe as a mere amateur could. It could be both frustrating and exciting. For every hundred dead-ends, one curious, fascinating, shiny thing would appear in my prospector's pan, some of which I have described — and perhaps claimed too much for — in this book.

§

Some of the material in this book has been published before. The proof that number-sequence problems invariably admit to an infinite number of solutions was first published in 1994, in issue 285 of *Tableaus* (the journal of Australian Mensa). In the latter 1990s I had a regular column in *Tableaus* called *Mathematical Doodlings*, and some of the doodlings published there are reprinted here. And some of the conjectures concerning prime numbers offered later in the book first appeared in 2005 in *Journal of Recreational Mathematics* as "Ten Prime Conjectures"[7]. Finally, some material is reproduced from my 1988 book *Think About It!*. But scattered within this harvest of old-growth material are some new doodlings, trifles perhaps, but hopefully an inspiration to others whose love of numbers, and keenness to explore, risks being overwhelmed by the impenetrable complexities that occupy professional mathematicians. Grant them their sandstone towers; but the sandpit of numbers is everyone's for playing in.

§

Finally, my thanks are owed to Marcia Bascombe for her expert editorial assistance.

---

7. Vol. 33, no. 3, 2004–2005, pp. 193–196.

$$(n) \sim \frac{\sqrt{2n}}{\log n} \cdot \prod_{w=3}^{\infty} \left(1 - \frac{1}{w^2 - 1} \cdot \left(\frac{2n}{w^2}\right)\right)$$

Part 1

Curiosities, conjectures and challenges

2, 4, 6, 8, 10, -3243: Cognitive disarray or rational answer?

Anyone who has sat a typical IQ or scholastic test will have encountered questions asking for the next number in a numerical sequence or for the numbers missing from a given portion of a numerical sequence. A simple example is this: {2, 4, 6, 8, 10, ?}, to which 12 springs to mind as the required answer. Many will have experienced the niggling doubt that there may well be more than one answer to such questions, a doubt likely to be well rooted if they had previously encountered IQ tests (or done number-sequence puzzles in puzzle books) and observed that a numerical sequence need not always be understood as a strict arithmetic series. It may, instead, be merely an ordered set of like numbers (for example, prime numbers) rather than a series of numbers that progress by some identifiable mathematical formula. And so, in the absence of any specified context or instruction on how to interpret the given sequence, many answers may fit the puzzle.

In addition, the manifold complexity of arithmetic itself easily mocks attempts to generate, in a strictly arithmetic manner, a unique number-sequence from just three or four given numbers. For example, consider the following part-sequence:

$$\{1, 2, 3, ?, ?, ? \ldots\}$$

How very natural it is to continue the series {... 4, 5, 6 ...} on the assumption that the given numbers are part of the most natural series of all: the series of integers that mathematicians call the natural numbers.[1] But suppose someone coming across such a part-sequence in an IQ test continues it {... 5, 8, 4 ...} Should this be construed as cognitive perversity and marked wrong? If the examining psychologist should think so, they could well be making a grave mistake. For the numbers can be construed as a series in which the sum of the preceding two numbers is reduced, each time, to its digital root. (The digital root of a number is the sum of the digits of that number, further reduced, if necessary, to a single digit by continual digit summing. Hence, the digital root of 35 is 3 + 5 = 8, and the digital root of 178 = 1 + 7 + 8 = 16 = 1 + 6 = 7.) Consequently, {... 5, 8, 4 ...} is a legitimate answer to the number-sequence puzzle { 1, ?, ?, ? ...}.

And so, too, is {... 5, 7, 11 ...} if the sequence is thought of as the ordered set of numbers that are divisible only by themselves and 1 (in other words, the prime numbers including 1).

And so, too, is {... 5, 9, 16 ...}, for the simple polynomial

$$y = \frac{1}{6}x^3 - x^2 + \frac{17}{6}x - 1$$

yields, for consecutive values of x starting with zero, the arithmetic series {1, 2, 3, 5, 9, 16, 27 ...}.

And so too is {... 5, 8, 13 ...} if the numbers are construed as a Fibonacci series with starting numbers 1 and 2 rather than 1 and 1. In other words, with the exception of 1 and 2, each number is the sum of the preceding two numbers.

Here, then, are five ways of continuing the sequence {1, 2, 3 ...}, and a little imagination should yield considerably more.

---

1. An integer is just a whole number: –3, –2, –1, 0, 1, 2, 3 etc.

Of course, offering just three starting numbers in a number-sequence puzzle is inviting a multiplicity of answers. Perhaps offering four will cut out the multiplicity. Well, it depends. Take {8, 16, 24, 32, ?, ?}, for example. One solution that might seem obvious is {8, 16, 24, 32, 40, 48 ...}. Here each digit is 8 times its place in the sequence. But what is wrong with {8, 16, 24, 32, 42, 50 ...}? Each number in the sequence (N) is obtained from the formula $N = p + 6n + 1$, where p is the value of an odd prime number (including 1) and n is the place of that prime number in the ordered set of all odd numbers.[2] Thus we have:

$$8 \ = \ 1 + (6 \times 1) + 1$$
$$16 \ = \ 3 + (6 \times 2) + 1$$
$$24 \ = \ 5 + (6 \times 3) + 1$$
$$32 \ = \ 7 + (6 \times 4) + 1$$
$$42 \ = \ 11 + (6 \times 5) + 1$$
$$50 \ = \ 13 + (6 \times 6) + 1$$

and so on.

Even the seemingly simple puzzle {?, 4, 6, 8, 10, 12, 14 ...} admits to more than one answer. You might swiftly decide that 2 is the right answer. But what is wrong with 0? For the sequence {0, 4, 6, 8, 10, 12, 14 ...} is simply the sequence of non-negative even integers that are not also prime numbers.

The elements in the mathematical universe seem, then, to be richly connected. For any number, there are multiple ways of getting to it from any other number (or

---

2. The integer 1 is no longer considered a prime number, even though it meets the condition that it is divisible only by itself and one. It once was. For example, Goldbach—whose views we consider later in this book—considered 1 to be a prime, as did many mathematicians in the Middle Ages and the Renaissance.

sequence of numbers). If only to prevent themselves from misjudging a truly creative intelligence, intelligence-testing psychologists might, then, do well to require an explanation to be given for why a number sequence that forms part of an IQ test is continued in a particular way. They don't do this. Instead they try to limit the number of possible answers to one by making the test item a multiple-choice question.

But the richness of the mathematical universe can thwart even the most diligent of test designers in their quest to ensure that only one answer is possible. To demonstrate this, consider the following question from a sample IQ test designed by IQ Test Labs.[3]

> 3. Please enter the missing figure: 4, 5, 8, 17, 44,
>
> A. 80
> B. 125
> C. 112
> D. 60
> E. 84

The answer given is B and the reasoning provided is this: "The difference between the numbers follows the series 1, 3, 9, 27, 81". Well, that's one reason for choosing B, but another is that each figure is three times the previous figure less 7. For example, $5 = (4 \times 3) - 7$; $8 = (5 \times 3) - 7$, $17 = (8 \times 3) - 7$; and so on. Another reason is that only B continues the sub-sequence created by subjecting each number to a mod 2 calculation, namely, $\{0, 1, 0, 1, 0 \ldots\}$.[4] In other words, only B continues the alternation of even and odd numbers. Another reason for choosing B is that the differences between the even numbers in the sequence are 4 and 36 and 36 is 4×9. The differences

---

3. See http://www.intelligencetest.com/questions/index.htm. Viewed 14 October 2016.

between the odd numbers are 12 and ... we're not sure yet, but if we assume the same factor between differences as for the even numbers (12), the next difference would be $9 \times 12 = 108$ and $17 + 108$ just so happens to be 125. Another reason: form two sub-sequences, one for the even numbers, the other for the odd. The differences between corresponding numbers in the two sub-sequences create the sequence $1^2 + 0$, $2^3 + 1$, $3^4 + 0$, $4^5 + 1$ and so on.

Thus the richness of the mathematical universe is clearly exposed. But can that richness yield answers other than B in the list of possible answers? It surely can. Let's start with answer C: 112. One pattern that can easily be spotted is that $(4 \times 8) + (4 + 8) = 44$, thereby linking the first, third and fifth numbers in the sequence. From here the rest is easy. Let n be the position of a number in the sequence (so that $4 = 1$, $5 = 2$, $8 = 3$, $17 = 4$ and so on). Further, let a be the first number, b the second, c the third, d the fourth, e the fifth and f the number that is missing. Now $(a \times c) + ((n \times a) + c) = e$ (that is, $(4 \times 8) + ((1 \times 4) + 8) = 44$). Let's apply the same formula to the next pair of alternate numbers: $(b \times d) + ((n \times b) + d) = f$. With the numbers provided we have $(5 \times 17) + ((2 \times 5) + 17) = 112 =$ answer C. QED. Another discernible pattern leading to answer C is this: multiply the first and fourth number and add the result to the fifth, multiply the second and fifth number and add it to the sixth, multiply the third and sixth and add it to the seventh and so on ad infinitum. The sequence so generated is {4, 5, 8, 17, 44, 112, 332, 428 ...}.

Now consider answer E: 84. If we again view the sequence as two sub-sequences of alternating numbers,

---

4. An a mod b calculation yields the remainder when a is divided by b. Thus 125 mod 2 yields 1, which continues the mod-2 sequence {0, 1, 0, 1, 0, 1 ...}. By the way, n mod 4 also yields the sequence {0, 1, 0, 1, 0, 1 ...}, giving another reason for choosing B.

we can derive 44 by omitting the first digit from $(a + c)^2$ and likewise 84 by omitting the first digit from $(b + d)^2$. Further, note that $(b + d) \times 2 = e$ — that is, $5 + 17 = 22 \times 2 = 44$) — and $(a + c) \times 7 = f$: 84. If the multipliers (2 and 7) repeat, then the sequence continues {4, 5, 8, 17, 44, 84, 202, 364 ...}. Note that if the multipliers are 2 and 5 rather than 2 and 7, we get the sequence {4, 5, 8, 17, 44, 60, 154, 260 ...}, which would satisfy answer D. We could also get answer E by firstly adding b and d, reversing the digits and multiplying by 2, and then adding a and c, reversing the digits and multiplying by 4. If the multipliers were to repeat, we get the sequence {4, 5, 8, 17, 44, 84, 100, 202 ...}.

Another way to get 60 (answer D) is to notice that $c = (a + b) - 1$, $d = (b + c) + 4$ and $e = (c + d) + 19$. Now if the constants were to repeat the pattern {–1, 4, 19, –1, 4, 19 ...}, then $60 = (d + e) - 1$ and the sequence would continue {4, 5, 8, 17, 44, 60, 108, 187 ...}.

So our sample IQ test question {4, 5, 8, 17, 44, ?} admits to more than one answer.

Let's return to the sequence given at the start of this section, namely {2, 4, 6, 8, 10, ?}. One answer, and the one that would no doubt be marked correct on a psychometric test, is 12. An answer of 13 would probably be interpreted as a hurried slip of the pen; an answer of –3243 is likely to be interpreted as evidence of cognitive confusion. Yet all three answers (12, 13 and –3243) are equally correct, correct in the sense that there is a knowable rule that will generate the sequence as continued.

In fact, as I am about to show, *any* number can continue the above sequence — and any of the sequences we have considered so far — and the sequence as continued will be the product of a strict mathematical rule (an equation, in other words — a polynomial, to be precise). The proof — which requires no sophisticated mathematical knowledge — follows from a curious feature of polynomials.

A polynomial is an algebraic expression of the general form:

$$y = ax^n + bx^{n-1} + cx^{n-2} + \dots + gx + h$$

One sort of polynomial you are sure to be familiar with — probably through years of dreary classroom factorisation — is the quadratic equation. An example is $y = x^2 + 2x + 3$. A quadratic equation is said to be a polynomial of degree 2, since 2 is the highest power in such an equation. Polynomials range from degree zero (for example, $y = 3$) to any positive degree you might wish to imagine. For our discussion it is important to note that a number sequence can be generated by repeatedly plugging consecutive x values into a polynomial and solving for y. Using the quadratic mentioned above, plugging in one value at a time from the set of consecutive x values $\{0, 1, 2, 3, 4, 5 \dots\}$ yields the number sequence $\{3, 6, 11, 18, 27, 38 \dots\}$.

Now to an interesting feature of many polynomials (and certainly all those up to degree 6): a sequence of numbers derived from a polynomial is reducible, by the repeated subtraction of adjacent pairs, to a zero sequence. Let me explain this using, as an example, the quadratic polynomial mentioned above, namely $y = x^2 + 2x + 3$. As we noted, the sequence of y values $\{3, 6, 11, 18, 27, 38 \dots\}$ can be derived by substituting consecutive integers for x in this equation starting with $x = 0$. Taking each pair of numbers in turn and subtracting the number on the left from the number on the right yields a further sequence: $\{3, 5, 7, 9, 11 \dots\}$. Repeating this procedure but this time using the figures in the newly generated sequence gives the sequence $\{2, 2, 2, 2 \dots\}$. Repeating the procedure again yields a zero sequence:

$\{0, 0, 0 \ldots\}$. This process of repeated subtraction can be displayed graphically like this:

$y =$	3		6		11		18		27		38
(1)		3		5		7		9		11	
(2)			2		2		2		2		
(3)				0		0		0			
(4)					0		0				
(5)						0					

where each number in a subtraction sequence—any of those labelled (1) to (5) at the left—is derived by the right-to-left subtraction of the two figures immediately above it. I will call this layout a *subtraction triangle*. Obviously a subtraction triangle can be generated for any sequence of numbers. Further, if the sequence is finite, it will run out of pairs after $(t - 1)$ subtraction sequences, where t is the number of figures given in the original sequence. (Note that 6 figures are given in the original sequence above and there are no more pairs after the fifth subtraction sequence.) I will call the last figure in any particular subtraction triangle—the one without a pair—the *concluding term*.

Just as we can derive the subtraction triangle for a sequence generated by a specific polynomial—as we just did—we can derive the subtraction triangle for the *general* sequence derivable from the *general* form of some polynomial. For example, the general form of polynomials of degree 2—quadratic equations, in other words—is $y = ax^2 + bx + c$. From this the subtraction triangle shown in Table 1 on page 31 can be derived.

This subtraction triangle shows that a sequence generated by a quadratic—any quadratic—yields, at the third level of subtraction, a zero sequence, and does so irrespective of how far one extends the sequence of given terms.

Table 1: Subtraction triangle for the general form of a quadratic equation: $y = ax^2 + bx + c$

$x =$	0	1	2	3	4	...
$y =$	c	$a+b+c$	$4a+2b+c$	$9a+3b+c$	$16a+4b+c$	...
(1)		$a+b$	$3a+b$	$5a+b$	$7a+b$	$[9a+b]$
(2)			$2a$	$2a$	$2a$	$[2a]$
(3)				0	0	$[0]$

Table 1: Subtraction triangle for the general form of a quadratic equation (continued)

$x =$	...	$n-3$	$n-2$	$n-1$	n
$y =$		$a(n-3)^2 + b(n-3) + c$	$a(n-2)^2 + b(n-2) + c$	$a(n-1)^2 + b(n-1) + c$	$an^2 + bn + c$
(1)			$2an - 5a + b$	$2an - 3a + b$	$2an - a + b$
(2)		$[2a]$	$2a$	$2a$	
(3)			0	0	

And now for a second interesting feature of the general subtraction triangle for quadratics shown on page 31. The starting terms of all the subtraction sequences down to the zero sequence ($a + b$ and $2a$) together with the starting term of the given sequence (c) together form, with the corresponding starting terms for a specific quadratic, a set of simultaneous equations that enable us to calculate a, b and c. In other words, we can, from a subtraction triangle that reaches a zero sequence at level 3, derive the coefficients of an equation that generates the given sequence and thus the equation itself. All we need do is consider the specific starting terms of each subtraction sequence.

For example, {3, 11, 25, 45, 71 ...} are consecutive y values that can be generated by a quadratic equation. You can tell this by deriving the first zero sequence at subtraction level 3:

$y =$	3		11		25		45		71
(1)		8		14		20		26	
(2)			6		6		6		
(3)				0		0			
(4)					0				

If we now superimpose our general template for quadratics — see page 31 — over the subtraction triangle above, we see that $c = 3$, $a + b = 8$, and $2a = 6$. So $a = 3$ making b equal to 5. Thus the generating quadratic is:

$$y = 3x^2 + 5x + 3$$

And you can check that this quadratic does generate the given sequence by plugging in consecutive x values beginning with 0 and simplifying the result. For example, when $x = 2$ (which is the third value in the sequence {0, 1, 2, 3 ...}, $y = (3 \times 2^2) + (5 \times 2) + 3 = 25$, and 25 is the third number in the given sequence.

General subtraction triangles can be derived in like fashion for polynomials of any degree. I have done so as far as degree 6 (after which the tedium that necessarily befalls the amateur doodler kicked in). Table 2 on page 34 gives, for all polynomials of degree 1 to 6, the starting terms for each subtraction sequence as far as the first zero sequence. I derived these equations by creating a subtraction triangle for the general form of each polynomial shown in the top row of the table, just as I did above for quadratics. And in so doing, I proved two things:

1. a zero sequence always appears first after $(n + 1)$ steps, where n is the degree of the polynomial and

2. the starting terms of the subtraction sequences as far as the first zero sequence—and including the first term of the given sequence when $x = 0$—are always sufficient to form a set of simultaneous equations that, in any instance, enable us to derive the coefficients of a polynomial that could generate the subtraction triangle and thus the polynomial itself.

In other words, what we did with {3, 11, 25, 45, 71 ...} to determine the quadratic that could generate it can be done with any sequence generated by a polynomial of degree 1 to 6. For example, consider the sequence: {4, 5, 8, 17, 44, 109 ...}. The associated subtraction triangle for this sequence reveals a zero sequence at subtraction level 5, with the subtraction sequences beginning 1, 2, 4 and 8 respectively:

$y =$	4	5	8	17	44	109
(1)		1	3	9	27	65
(2)			2	6	18	38
(3)				4	12	20
(4)					8	8
(5)						0

Table 2: Template of starting terms for subtraction triangles of polynomials of degree 1–6

$y =$	$ax + b$	$ax^2 + bx + c$	$ax^3 + bx^2 + cx + d$	$ax^4 + bx^3 + cx^2 + dx + e$	$ax^5 + bx^4 + cx^3 + dx^2 + ex + f$	$ax^6 + bx^5 + cx^4 + dx^3 + ex^2 + fx + g$
Last coefficient =	b	c	d	e	f	g
(1)	a	$a + b$	$a + b + c$	$a + b + c + d$	$a + b + c + d + e$	$a + b + c + d + e + f$
(2)		$2a$	$6a + 2b$	$14a + 6b + 2c$	$30a + 14b + 6c + 2d$	$62a + 30b + 14c + 6d + 2e$
(3)			$6a$	$36a + 6b$	$150a + 36b + 6c$	$540a + 150b + 36c + 6d$
(4)				$24a$	$240a + 24b$	$1560a + 240b + 24c$
(5)					$120a$	$1800a + 120b$
(6)						$720a$

Hence a polynomial of degree 4 can generate this sequence. The equations for the starting terms of the subtraction sequences derived from polynomials of degree 4—given in the fifth column of Table 2 on page 34—allow us to conclude that $24a = 8$, $36a + 6b = 4$, $14a + 6b + 2c = 2$, $a + b + c + d = 1$ and $e =$ the first term in the given sequence. We now have a set of simple simultaneous equations to solve which shows us that a possible generating polynomial is:

$$y = \frac{1}{3} \cdot x^4 - \frac{4}{3} \cdot x^3 + \frac{8}{3} \cdot x^2 - \frac{2}{3} \cdot x + 4$$

And to prove that this polynomial does generate the sequence {4, 5, 8, 17, 44, 109 ...}, just plug the x values {0, 1, 2, 3, 4, 5 ...} in turn into the polynomial and simplify the result.

Given that the approach I took in generating Table 2 on page 34 was exactly the same with each degree of polynomial—derive subtraction sequences from the sequence generated by plugging the values 0, 1, 2, 3 ... into the general form of the polynomial—I am prepared to conjecture that discoveries 1 and 2 noted on page 33 apply to *all* polynomials, irrespective of degree.

Now to a proof that an infinite number of polynomials can generate the sequences found on typical IQ tests. If we are successful here, and if discoveries 1 and 2 noted on page 33 apply to all polynomials, then we can be justified in claiming that *any* finite sequence can be continued in an infinite number of ways and each continuation can be generated by a polynomial. But let's start modestly, with sequences that have just a few starting terms. To prove my point, I'll derive the subtraction triangle corresponding to the sequence {1, 4, 9, 16 ...}, although any similar sequence will give the very same conclusion.

$y =$	1		4		9		16		P
(1)		3		5		7		$P-16$	
(2)			2		2		$P-23$		
(3)				0		Q			

The first thing to note in this triangle is that there is a subtraction sequence that could be continued as a zero sequence, namely, sequence (3). Hence we are entitled to assume that there is a polynomial of degree 2 that could generate the given y sequence. Indeed, using our template for polynomials of degree 2 — given in Table 2 on page 34 — we can quickly derive the form of this polynomial, namely, $y = x^2 + 2x + 1$. We can check that this equation does indeed yield the given y sequence by plugging in consecutive values for x starting from zero and solving for y. We could also use this equation to calculate a possible next term in the sequence: 25.

It is important to understand why we are entitled to assume that subtraction sequence (3) is a zero sequence even though it has only one number in it. We can do so because the only constraint we are under in creating subtraction sequences is to ensure that together they generate the numbers in the initial sequence and in the given order. And the numbers in {1, 4, 9, 16 …}, and their order, is in no way affected by our assuming that subtraction sequence (3) continues {0, 0, 0 …}. In other words, we are entitled to extend and change a subtraction triangle in any we please *so long as the original y sequence remains unchanged.*

Now just as we are entitled to assume that subtraction sequence (3) in the above subtraction triangle is a zero sequence, we are entitled to assume that it is *not.*

Given the numbers we have in the y sequence, Q—the next term in sequence (3)—can be any value at all. In other words, whatever value we give to Q, the part-sequence given—{1, 4, 9, 16 ...}—will not change.

Bearing this in mind, let's shift our attention from Q to P, the next term in the given y sequence. (In the typical number-sequence problem on an intelligence test, P is the number being asked for.) We have already found a polynomial to generate a value for P—namely, 25—but let's give P some other value. Any value, in fact: positive or negative. What we now find is that by continuing the subtraction triangle—creating lower-level subtraction sequences, in other words—another subtraction sequence can be found—or created—that is a zero sequence. And this means that there must be another polynomial that can create the particular y sequence we are investigating. Let's see how this works by substituting, say, 13 for P in the y sequence under consideration and continuing the subtraction triangle:

$y =$	1		4		9		16		13
(1)		3		5		7		–3	
(2)			2		2		–10		
(3)				0		–12			
(4)					–12		R		
(5)						S			

Our extended subtraction triangle runs out of puff at subtraction level (4), there being no partner for the new concluding term (–12) to be subtracted from. But there is no reason why we cannot continue subtraction sequence (4) and do so as we please, since doing so will not affect the given y sequence {1, 4, 9, 16 ...}. So let's make R equal to the current concluding term: –12. This enables us to create subtraction sequence (5), with S, our new

concluding term, equal to zero. Because S is equal to zero
we are entitled to assume that subtraction sequence (5) is
the start of a zero sequence. Hence there must also be a
polynomial of degree 4 (that is, the number of the
subtraction level corresponding to a zero sequence less 1)
capable of generating a sequence starting $\{1, 4, 9, 16 \ldots\}$.
This is in addition to the quadratic we uncovered earlier
when we stopped at subtraction level 3. We can now
apply the template for polynomials of degree 4 (shown in
Table 2 on page 34) to derive the equation for this
polynomial, which turns out to be:

$$y = -\frac{1}{2}x^4 + 3x^3 - \frac{9}{2}x^2 + 5x + 1$$

You can check that this equation does generate the
sequence $\{1, 4, 9, 16, 13 \ldots\}$ by plugging in consecutive x
values beginning from zero and solving for y.

Of course, we didn't have to make $P = 13$. We picked a
number just to show that there will, or can, be a zero
sequence in the subtraction triangle, and a zero sequence
always indicates that a polynomial can generate the top-
level sequence. But the number we do make P will partly
determine the value that starts subtraction sequence (4),
and this term will partly determine the coefficients of the
corresponding polynomial. So there will be a unique
polynomial for each possible value of P. Since P can be
any number at all and a zero sequence found, there are
any number of polynomials of degree 4 capable of
generating the sequence $\{1, 4, 9, 16, ?\}$. *An infinity in fact.*

But the infinity of possibilities so far uncovered is but
a drop in the ocean. For each additional term with which
we continue the given y sequence, there will be yet
another infinity of polynomials capable of generating that
sequence. For the term that follows P in the above

sequence enables the subtraction triangle to be extended by one additional subtraction level. In other words, where P caused the concluding term to fall at the start of subtraction sequence (4), the term after P will cause the concluding term to fall at the start of subtraction sequence (5). Again we can continue the final subtraction sequence with identical terms, thus enabling us to create a further subtraction level — subtraction sequence (6) — which can be the start of a zero sequence. This means that there must be polynomials of degree 5 capable of generating the given y sequence. Let's prove this by continuing the sequence $\{1, 4, 9, 16, 13, 22 \ldots\}$. The corresponding subtraction triangle is:

$y =$	1		4		9		16		13		22
(1)		3		5		7		–3		9	
(2)			2		2		–10		12		
(3)				0		–12		22			
(4)					–12		34				
(5)						46		T			

If we make T equal to 46 — which doesn't affect the original sequence — we will have the start of what could be a zero sequence at level 6. Plugging the starting terms (1, 3, 2, 0, –12 and 46) into the degree-5 template in Table 2 on page 34 and solving the resulting simultaneous equations yields the following polynomial:

$$y = \frac{23}{60} \cdot x^5 - \frac{13}{3} \cdot x^4 + \frac{197}{12} \cdot x^3 - \frac{71}{3} \cdot x^2 + \frac{71}{5} \cdot x + 1$$

This generates the sequence $\{1, 4, 9, 16, 13, 22 \ldots\}$ for consecutive values of x starting with 0.

This is, of course, one of an infinity of polynomials of degree 5: one for each possible value that the term after P can take.

And then, of course, there is the term after the term after P — generating an infinity of polynomials of degree 6. And so on, possibly *ad infinitum*.

To summarise what we have found: if there are 4 numbers in a given finite

A challenge: We have seen that polynomials of power 2, 4, 5 and 6 can generate a plausible continuation of the sequence {1, 4, 9, 16 ...}. Why are polynomials of power 3 excluded?

The answer: page 185.

sequence, there will be an infinity of fourth-degree polynomials that can generate the sequence as given, one for each possible fifth value. If there are 5 numbers in a given finite sequence, there will be an infinity of fifth-degree polynomials that can generate the sequence as given, one for each possible sixth value. If there are 6 numbers in a given finite sequence, there will be an infinity of sixth-degree polynomials that can generate the sequence as given, one for each possible seventh value. And thus we conjecture that if there are n numbers in a given finite sequence, there will be an infinity of nth-degree polynomials that can generate the sequence as given, one for each possible ($n + 1$) value. Moreover, any given sequence that can be generated by an infinity of nth-degree can also be generated by an infinity of polynomials of degree ($n + 1$). In other words, we have what might be called *latitudinal infinity* (that is, infinity within a degree) and *longitudinal infinity* (that is, infinity across degrees).[5]

The above technique can be used to uncover generating polynomials for any sequence, no matter how seemingly unruly or random the numbers in it appear. Let's illustrate this using an unlikely sequence of numbers, a sequence I'll

---

5. One is tempted to speculate that a polynomial might be capable of taming that most fascinating sequence of numbers, namely, the prime numbers.

just pull out of the air: {3, –112, 69, 1003, –47 ...}. The corresponding subtraction triangle looks like this:

$y =$	3		–112		69		1003		–47
(1)		–115		181		934		–1050	
(2)			296		753		–1984		
(3)				457		–2737			
(4)					–3194		X		
(5)						Y			

Without it affecting the given sequence, we can substitute –3194 for X and get a value of zero for Y. Hence we can treat sequence (5) as the beginning of a zero sequence. Thus we know that some polynomial can generate the given y sequence. We also know that it is a polynomial of degree 4—since subtraction level (5) is a possible zero sequence—and so we know the appropriate template to apply to derive the particular equation. Applying that template gives us the following polynomial:

$$y = -\frac{1597}{12}x^4 + \frac{2624}{3}x^3 - \frac{18533}{12}x^2 + \frac{4127}{6}x + 3$$

Plugging in consecutive values for x from the set {0, 1, 2, 3, 4 ...} will verify that this polynomial does indeed generate the sequence {3, –112, 69, 1003, –47 ...}. If we had made X in the subtraction triangle some value other than –3194, we could have continued subtraction sequence (5) by repeating whatever value Y had become. Then we could create a zero sequence as subtraction sequence (6) and derive a polynomial of degree 5. In fact, we could derive a unique polynomial of degree 5 for each possible X value.

It might be argued that I have fallen foul of the fallacy of affirming the consequent. I have argued that a sequence derived from a polynomial generates a zero sequence in a subtraction triangle, and then argued that if a zero sequence appears in a subtraction triangle, the generating sequence is derived from a polynomial. Laid out that way, my argument is this: if *A* then *B*, *B* hence *A*. And this is indeed the fallacy of affirming the consequent. An example: if it is raining, the ground is wet; the ground is wet; therefore it is raining. No, there's a gardener outside with a hose!

Let me counter this in two ways. Firstly, I am not arguing that a zero sequence indicates that *only* a polynomial can generate the starting sequence. That would clearly be wrong. There are many types of equation other than polynomial — exponential, trigonometric, logarithmic and so on — each of which generates a sequence of numbers for consecutive vales of the independent variable. Secondly, as my table of poly-nomial templates shows (see page 34), a subtraction triangle generated by a polynomial always gives us a set of simultaneous equations that enables us, for any specific subtraction triangle, to derive a polynomial that *could* generate the given sequence. And all I set out to show is that a polynomial *could* generate any given finite sequence — *regardless of how it is continued.*

Let's return to the sequence that opened this paper: {2, 4, 6, 8, 10, ?...}. We postulated that an answer of 13 might generously be interpreted as a hurried slip of the pen while an answer of −3243 will more likely be seen as sullen rejection or cognitive impairment. And yet both can be derived from humble polynomials. For example, if we continue the sequence with 13 and derive the associated subtraction triangle, we find a concluding term of 1 at subtraction level (5). Hence a zero sequence can be created at subtraction level (6), indicating the presence of a generating polynomial of degree 5. Our template for such polynomials gives us its exact form:

$$y = \frac{1}{120}x^5 - \frac{1}{12}x^4 + \frac{7}{24}x^3 - \frac{5}{12}x^2 + \frac{11}{5}x + 2$$

The sequence generated by this polynomial is {2, 4, 6, 8, 10, 13, 20, 37, 74, 146 ...}.

And what about that seemingly perverse answer −3243? Well, here is the formula:

$$y = -\frac{217}{8}x^5 + \frac{1085}{4}x^4 - \frac{7595}{8}x^3 + \frac{5425}{4}x^2 - 649x + 2$$

My conjecture that all finite sequences can be continued in any way at all and the continuation will be the product of a polynomial began with an examination of polynomials of degrees 1 to 6. I showed that for sequences generated by these polynomials, a zero sequence is always generated in the associated subtraction triangle and that the starting terms of the subtraction sequences always enable us to determine the exact form of the generating polynomial. Apart from the tedium of pen-and-paper algebra, there is another reason why I stopped at degree 6. What I am especially interested in is the number-sequence questions that appear on intelligence tests. We noted earlier that a subtraction triangle will run out of pairs after $(t - 1)$ subtraction sequences, where t is the number of figures given in the original sequence. The solitary digit remaining we called the *concluding term*. Very few number-sequence questions on intelligence tests offer more than 7 numbers as starting terms — most offer less — which means that the subtraction triangle will run out of pairs after 6 subtraction sequences. If a zero sequence hasn't been encountered in the subtraction triangle, we can repeat the concluding term, thus forcing a zero sequence at the seventh subtraction sequence. And a zero sequence at the seventh subtraction sequence entails that a polynomial of degree 6 can generate it. So, by stopping at polynomials of order 6, I will have met my goal of casting doubt on the validity of number-sequence questions that appear on intelligence tests.

The obvious question to pose is this: given that number-sequence problems feature prominently in IQ and other psychometric tests, what reasons are there for marking any answer to a number-sequence problem as wrong? If a problem admits to an infinity of equally correct answers and yet provides no objective way of ranking the possible answers in a way that assists the measuring of what the problem was designed to measure then surely it cannot be a valid test item. Consequently, such problems should, perhaps, be banished from intelligence tests. That is a point I will take up in detail in appendix A, "The pitfalls in measuring intelligence" starting on page 141.

Let's revisit the IQ test question dissected on page 26.

3. Please enter the missing figure: 4, 5, 8, 17, 44,

A. 80
B. 125
C. 112
D. 60
E. 84

The answer given is B. But as we've just shown, every answer matches a pattern generated by some polynomial — indeed, by an infinite number of polynomials. And here is just one for each of the supposedly non-correct answers to {4, 5, 8, 17, 44, ? }:

Answer A (80) will be generated by:

$$\frac{-29}{120} \cdot x^5 - \frac{50}{24} \cdot x^4 + \frac{461}{24} \cdot x^3 - \frac{461}{12} \cdot x^2 + \frac{338}{15} \cdot x + 4$$

Answer C (112) will be generated by:

$$\frac{1}{40} \cdot x^5 + \frac{1}{12} \cdot x^4 - \frac{11}{24} \cdot x^3 + \frac{17}{12} \cdot x^2 - \frac{1}{15} \cdot x + 4$$

Answer D (60) will be generated by:

$$\frac{-49}{120} \cdot x^5 + \frac{53}{12} \cdot x^4 - \frac{375}{24} \cdot x^3 + \frac{277}{12} \cdot x^2 - \frac{157}{15} \cdot x + 4$$

Answer E (84) will be generated by:

$$\frac{-5}{24} \cdot x^5 + \frac{29}{12} \cdot x^4 - \frac{207}{24} \cdot x^3 + \frac{157}{12} \cdot x^2 - \frac{17}{3} \cdot x + 4$$

None of these polynomials produce bizarre or unruly graphs. The graph below is a visual representation of the polynomial immediately above (for answer E). Most secondary school students would have encountered graphs like this is the classroom. So it is not as if giving polynomials the task of continuing sequences requires us to enter the arcane and daunting world of tertiary mathematics.

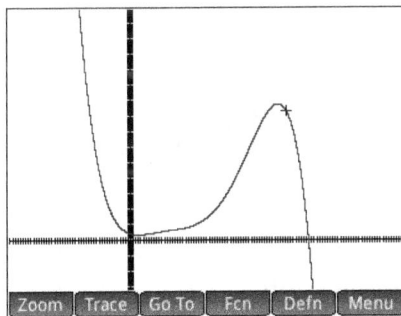

A proponent of IQ tests might retort than we are barking up the wrong tree here, that number-sequence questions are not testing mathematical prowess but mere pattern recognition. In that case, what is wrong with the multiplicity of answers given on pages 27 and 28. None of those answers required any special mathematical prowess. Anyway, number-sequence questions are likely to be interpreted, by their very appearance, as requiring a mathematical answer. No context is given that will help the subject ascertain that all that the questioner is looking for is a pattern, not a mathematical equation.

Perhaps the retort will be made that these questions are looking for the *simplest* or most *obvious* answer (even though the question doesn't state that this is what is being tested). We'll come back to the fraught issue of what might be meant by *simple* and *obvious* in appendix A.

In some psychometric tests, pattern recognition is tested in ways other than with numerical sequences. Consider the following question from an intelligence test[6]:

2. Find the picture that follows logically from the diagrams to the right.

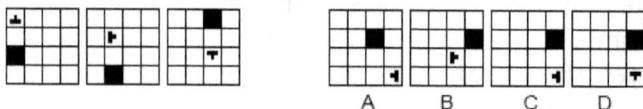

The answer given is C. You can reach C is if the following algorithm is applied to each frame to generate the next:

- Each object—the perpendicularity symbol and the black block—is moved one column to the right.
- Each object is moved down one row.
- If such a movement would take an object below the grid it is placed in the top row, and if such a movement would take an object to the right of the grid it is placed in the first column.
- Finally, each object is rotated 90°clockwise.

But again here are other answers (though clearly not an infinite number in this case). It is quite simple to devise

---

6. This question is taken from a sample intelligence test published by IQ Test Labs. See http://www.intelligencetest.com/questions/precognition.htm. Viewed 14 February 2016.

algorithms that would generate each one of the multiple-choice answers, a fact that renders worthless the use of this pattern-matching question on an intelligence test. I'll let you have a go at this yourself. Answers can be found in "Patterns before your eyes" on page 187.

So your niggling suspicion that there may be more than one answer to a number-sequence problem — or pattern recognition problem — on an IQ test is well founded.

Unusual multiplications

For my twelfth birthday my mother bought me a copy of *The Trachtenberg Speed System of Basic Mathematics*. This fascinating book is packed with techniques for mental arithmetic, techniques quite unlike any that were taught to me in the classroom. Trachtenberg's life was as fascinating as his mathematics. He began work in Russia, as an engineer. Following the Russian Revolution, he fell out with some of its leading lights. In 1919, fearing persecution, he fled on foot to Germany. The politics of liberalism that had got him into trouble with the revolutionaries of his homeland continued to occupy him and, in the 1930s, as he spoke out against fascism, he fell out with Hitler. Sensing his arrest, Trachtenberg fled to Vienna, only to be arrested later when Hitler's armies marched into Austria. He was imprisoned in a concentration camp where — with pestilence, starvation, stench and death all around him — he retreated into the world of numbers, devising the fascinating system of arithmetic shortcuts that now goes by his name. On learning of his impending execution, Trachtenberg's wife sold her remaining belongings and succeeded in bribing the camp officials into having her husband transferred to another camp, near Leipzig. From there he escaped, but was soon recaptured. This time he was sent to a camp in Trieste and sent to work breaking rocks. Again he escaped and, with his wife, managed to cross the border into Switzerland. In 1950, he founded the Mathematical Institute in Zurich, Switzerland—locally called the School of Genius—where he taught his system of mathematics. This brilliant, courageous man died three years later.

One of Trachtenberg's tricks (or shortcuts) that impressed me as a youngster was his way of squaring numbers ending in 5. He discovered that a quick way to do this was to use this equation:

$$(n \text{ attached to } 5)^2 = n(n + 1) \text{ attached to } 25$$

In other words, multiply the number formed by leaving off the 5 by the number one greater than it and append 25 to the result. Thus $35^2 = 3 \times 4$ with 25 added at the end, which is 1225. Likewise, $45^2 = 2025$ and so on.

Some years later, while idly doodling with numbers, I stumbled on a multiplication trick that incorporates the Trachtenberg squaring technique discussed above. Suppose you need to multiply a number ending in 5 (say 35) by a number 20 greater than it (in this case, 55). All you need do is square the number midway between the two (45)—and Trachtenberg's squaring technique is useful here—and then subtract 100 from the result. For example:

$$35 \times 55 = 45^2 - 100 = [4 \times 5]25 = 2025 - 100 = 1925$$

all of which can be done in your head. Another example:

$$105 \times 125 = 115^2 - 100 = [11 \times 12]25 = 13{,}225 - 100 = 13{,}125$$

And then curiosity got the better of me. I soon found numerous variations on the theme. For example, if you need to multiply a number ending in 5 (say 35) by a number 10 greater than it (in this case, 45), square the number midway between the two (40) and deduct 25. Thus:

$$35 \times 45 = 40^2 - 25 = 1600 - 25 = 1575$$

Further, suppose you need to multiply a number ending in 5 (say 35) by a number 30 greater than it. It turns out that all you need do is square the number midway between the two (50) and deduct 225. Thus:

$$35 \times 65 = 50^2 - 225 = 2500 - 225 = 2275$$

And if you need to multiply a number ending in 5 (say 35) by a number 40 greater than it, square the number midway between the two (55) and deduct 400. Thus:

$$35 \times 75 = 55^2 - 400 = 3025 - 400 = 2625$$

Expressing our results in tabular format will help us discern a pattern:

Difference (Δ)	Number to subtract
10	25
20	100
30	225
40	400

The number to subtract clearly progresses according to the formula:

$$\frac{\Delta^2}{4}$$

In other words, square the difference between the two 5-ending numbers and divide by 4. The result is the number you subtract from the square of the number midway between the two 5-ending numbers. And it is an algorithm that appears to apply whatever two 5-ending numbers you want to multiply.

Let's give it a try by multiplying 45 by 125:

$$45 \times 125 = 85^2 - \frac{80^2}{4} = 7225 - 1600 = 5625$$

That works. One more:

$$85 \times 145 = 115^2 - \frac{60^2}{4} = 13,225 - 900 = 12,325$$

And another for good measure:

$$65 \times 305 = 185^2 - \frac{240^2}{4} = 34,225 - 14,400 = 19,825$$

But now we are moving beyond the realm of feasible mental arithmetic, and away from the simplicity of Trachtenberg.

Common multiplications

You might be wondering what is so special about numbers ending in 5 that delivers them a unique multiplication technique. The answer is: there's nothing special at all. In fact, *any* two numbers can be multiplied using the technique discussed in the previous section, a technique expressed by the following equation:

$$p \times p = \left(\frac{p+q}{2}\right)^2 - \frac{(p-q)^2}{4}$$

And you can prove this by simplifying the right-hand side of this equation and noting that the simplest form of it is identical to the left-hand side:

$$= \frac{(p+q)^2}{4} - \frac{(p-q)^2}{4}$$

$$= \frac{(p^2 + 2pq + q^2) - (p^2 - 2pq + q^2)}{4}$$

$$= \frac{4pq}{4}$$

$$= p \times q$$

Since p and q can be *any* numbers, it follows that our technique is not limited to numbers ending in 5.

Let's instantiate this with an example, say, 24 × 36. By the equation above, we have:

$$24 \times 36 = \left(\frac{24 + 36}{2}\right)^2 - \frac{(24 - 36)^2}{4}$$

A spot of mental arithmetic—or a few taps on a calculator—will show that both sides of the equation yield identical answers: 864.

Infinite multiplications

The proof of the new multiplication technique offered in the previous section should tickle your curiosity about whether there might be other novel multiplication techniques waiting to be discovered. Here is one:

1. Square the sum of the two numbers being multiplied.
2. Square the difference between the two numbers being multiplied.
3. Calculate the difference between the two squares.
4. Divide by 4.

Here is a simple example that instantiates this technique. To multiply 28 by 22:

1. Square the sum: $50^2 = 2500$.
2. Square the difference: $6^2 = 36$.
3. Calculate the difference between the squares: $2500 - 36 = 2464$.
4. Divide the result by 4, giving 616.

One more for good measure:

$$251 \times 43 = (294^2 - 208^2)/4 = 10{,}793.$$

We can prove the generality of this method in much the same way as we proved the generality of the method explored in the previous section. Our new method can be expressed as:

$$= \frac{(p + q)^2 - (q - p)^2}{4}$$

By simplifying, we get:

$$= \frac{(p^2 + 2pq + q^2) - (q^2 - 2pq + q^2)}{4}$$

$$= \frac{4pq}{4}$$

$$= p \times q$$

This prompts us to ask whether there is a limit to the number of equations we could concoct that, when simplified, reduce to $p \times q$, which in turn prompts the following:

> **Conjecture**: there is an infinite number of equations each of which is equivalent to the multiplication of two numbers.

Old-fashioned multiplication. And new.

It is well known that you can multiply two numbers together using nothing more sophisticated than multiplication by 2, division by 2 and simple addition. Suppose, for example, that you needed to multiply 92 by 41. Taking the smaller of the two numbers (which is not strictly necessary, though it does reduce the overall number of steps) divide it by 2 and ignore the remainder (which yields 20 in our example). Divide this new number by 2 and keep doing likewise until you get to 1 (ignoring any remainder at each step). Now pair each number you derived with a number from the series $\{b, 2b, 4b, 8b, 16b ....\}$ where b is the larger of the two numbers you started with. In our example, the paired results look like this:

41	×	92
20		184
10		368
5		736
2		1472
1		2944

To find the product of 41 and 92, simply add up those numbers in the second column that are paired with an *odd* number in the first column: 92 + 736 + 2944 = 3772. Admittedly, this is a long-winded way to do multiplication compared to modern practices, but it is how multiplication was done at some time in the past.

It appears that a parallel technique to the one just described also works, parallel in that instead of taking the integer of each number you divide by two, you take the integer if there is no remainder and round up to the next integer (that is, next whole number) if there is a remainder. In this case, our table of results in the multiplication of 41 and 92 would look like this:

41	×	92
21		184
11		368
6		736
3		1472
2		2944
1		5888

You now add together all the numbers in the second column that are paired with an *even* number in the first column and then add the larger of your two starting numbers to the result: 736 + 2944 + 92 = 3772.

A feature of these types of multiplication tables is that the difference between the differences of all the odd pairs and the sum of all the even pairs is equal to the larger of the two numbers you started with. In our first example (see page 53), the equality is:

$$(2944 - 736 - 92) - (184 + 368 + 1472) = 92$$

This relationship holds whether you derive the numbers in the first column by integer division or by rounding-up division. In our example of rounding-up division, the equality is:

$$(5888 - 1472 - 368 - 184 - 92) - (736 + 2944) = 92$$

§

As we've seen, there are numerous ways to multiply numbers. As you might have expected, Trachtenberg too came up with a number of novel techniques. You can find these techniques described in *The Trachtenberg Speed System of Basic Mathematics* (Souvenir Press, London, 1962, translated and adapted by Anne Cutler and Rudolph McShane).

Special roots

Trachtenberg's neat technique for squaring numbers ending in 5 can be reverse-engineered to see if there are integer roots of a number ending in 25. By *integer roots* we mean a whole number such that, when it is squared, yields the starting number. For instance, 625 has an integer root of 25, since 25 × 25 = 625. (And since roots always come in pairs, 625 has another integer root, namely –25.)

Note that if a number ends in 5, its integer roots, if it has any, must also end in 5 (since no number other than 5 yields a number ending in 5 when squared).

And now to a technique for determining if there are integer roots of a number ending in 25. First, consider the digits other than the trailing 25. (For example, if the number we are investigating is 3025, consider just the digits 30.) If this truncated number does not end in 0, 2 or 6, you can conclude immediately that the full number does not have integer roots. Why? Recall that a 5-ending number squared is equal to the truncated number — the number without the 5 — multiplied by the number one greater than it. This is Trachtenberg's insight we discussed earlier. Thus the number before the 25 must end with a number that is the product of 0 and 1, 1 and 2, 2 and 3, 3 and 4 and so on. The only possibilities here are numbers ending in 0, 2 or 6. Thus you can tell by sight alone that 12,354,725 has no integer roots, since the digit before the trailing 25 is not 0, 2 or 6. It is 7.

Now find the largest integer that, when squared, is closest to, but not greater than, the truncated number. In the example of 30, 5^2 is 25 but 6^2 is 36. So the largest integer that, when squared, gives a value that is closest to, but not greater than, 30 is 5. Call this number the *root head* (the significance of which will become clear in a moment)[7].

Next deduct the square of the root head (25) from the truncated number (30). If you get the same number as the root head, then the full number has integer roots. In our example, 30 – 25 = 5, and 5 is the root head. Thus 3025 has integer roots. And the integer roots happen to be plus and minus the root head concatenated with 5. In our example, the roots are ±55. And indeed 55^2 and -55^2 yield 3025.

One more example: consider 140,625, of which the truncated digits are 1406. The largest integer that, when squared, is closest to, but not greater than, 1406 is 37 (since 37^2 is 1369 but 38^2 is 1444). So the root head is 37. Now deduct the square of the root head (1369) from the truncated number: 1406 – 1369 = 37. Note that the result is identical to the root head. Thus 140,625 has integer roots. And the integer roots happen to be ± the root head concatenated with 5: ±375. And indeed 375^2 and -375^2 = 140,625.

What if the truncated number—the number before the trailing 25—is a perfect square (that is, the product of two identical integers)? An example is 3625, since 36 = 6^2. In these cases we can tell straight away that the complete number does not have integer roots. Why? Let t be the

---

7. In mathematics, the largest integer that, when squared, gives a value that is closest to, but not greater than, another number, say p, is called the *integer square root* of p.

truncated number that happens to be a square (36 in our example) and u be the square root of t. Now t must, by Trachtenburg's insight, be equal to $s(s + 1)$, from which it follows that:

$$u = \sqrt{s \cdot (s + 1)}$$

Since u is an integer (6 in our example), $s(s + 1)$ must be a perfect square. In other words, there must be two consecutive integers such that their product must be a perfect square. But it should be clear that if you multiply two *consecutive* integers, the square root of the product must lie *between* the two integers. And since the two integers are consecutive, that square root cannot itself be an integer. Thus u in the above equation cannot be an integer, from which it follows that t cannot be a perfect square. To put it another way, if the truncated number is a perfect square, the 25-ending number cannot have integer roots. So 3625 does not have integer roots. (A few calculator taps shows that the square root of 3625 is approximately 60.208, which is a fractional number and not an integer.)

To conclude: if the truncated number does not end in 0, 2 or 6, or is a perfect square, you can tell immediately that the full number ending in 25 has no integer roots. In all other cases, the *root head* method described above can be used to check if the number has integer roots, and to find those roots if it has any.

More fun with numbers ending in 25

Consider again the case, described in the previous section, where the truncated number — the number before the trailing 25 — is a perfect square. The example we considered earlier was 3625, where the truncated number is 36. Take the square root of the truncated number and append a 5 to the result. Call this value r. Now calculate r^2. In our example, we get $65^2 = 4225$. Jot down the difference between this number and our starting number: 4225 − 3625 = 600.

Next square $r − 10$. In our example, we get $55^2 = 3025$. Notice that the difference between this number and our starting number is also 600.

Take another example: 12,125. Its truncated number is a perfect square: 11^2. Append a 5 and square the new number: $115^2 = 13,225$. Note the difference between this and the starting number: 13,225 − 12,125 = 1100. Now square 105. The answer − 11,025 − also happens to be different from the starting number by 1100.

> **Conjecture**: this symmetry holds whenever a number ending in 25 has leading digits that together form a perfect square.

Reversal arithmetic

Here is a simple trick that relies on features of a particular sort of subtraction that are not immediately obvious. Ask someone to write down − out of your view − a three-digit number. Next, ask them to create a new three-digit number by reversing the digits in the number they chose. (If, for instance, they chose 385, the new number will be 583.) Now ask them to subtract the smaller three-digit number from the larger three-digit number, that is, find the *absolute difference* between them. (In our example, they would subtract 385 from 583). When they have done that, ask them to create a new number by reversing the digits in the result of their subtraction. (To continue our example: 583 − 385 = 198; so our new number would be 891.) Then ask them to subtract the smaller of the last two numbers—the result of the earlier subtraction or the reversal of this result—from the larger of these two numbers. (For example, subtract 198 from 891.) Ask them to keep doing likewise until they come to zero, *as they always will.*

When your subject has reached zero, ask them to tell you the original three-digit number they chose. In a flash, you tell them the result of each subtraction in the string of subtractions that ended with zero. For instance: your subject says "952" and you instantly reply "693, 297, 495, 99 and zero". They check their subtractions and—lo and behold—you are right! So how is it done?

There are some very interesting features of three-digit reverse subtraction: each subtraction result (or difference) is exactly divisible by 99, and the result of repeated (or iterative) subtraction is always, eventually, zero.

But what makes this particular trick very easy to master are the following three special features of three-digit reverse subtraction: (1) the first digit in the result is always one less than the absolute difference between the first and the last digit of the number you are starting with; (2) the middle term is always 9; and (3) the last digit is always the difference between 9 and the number you arrived at in step 1.

Starting number:	693
Reverse:	396
Absolute difference:	297
Reverse:	792
Absolute difference:	495
Reverse:	594
Absolute difference:	99
Reverse:	99
Absolute difference:	0

For instance, if you are given 693 as the starting number, the first difference is 297, since, by (1), the first term is one less than 6 minus 3, the middle number is, by rule (2), always 9, and the third digit is 9 less the first digit (or 9 – 2 = 7). By applying the same simple rules to 297, you quickly arrive at 495. Applying the rules again brings you to 99 which, being a palindrome, is identical to its reversal and so, necessarily, generates zero when subtracted from its reversal.

Since these three rules are very simple (and only two require calculation, and of the simplest sort), you should be able to master this trick in a matter of minutes.

§

We noted above that if you continually reverse-subtract a three-digit number using as your starting term each time the result from the previous subtraction, you would always come down to zero. What might the result be if instead of subtracting mirror-image pairs, you added them?

Again, the result depends on the number of digits in the number you start with. But in the case of reverse addition, the exceptions are few—exceptions, that is, to the eventual arrival at a *palindrome*. (A palindrome is a number that reads the same from left to right as from right to left, such as 234432). For instance, if you start with 189, reverse it (yielding 981), add the two numbers (1170), reverse the result (711), add the result to its reversal, you get 1881, a palindrome).

The numbers 1 to 99 all result in a palindrome (and, indeed, in numerous palindromes as the reverse additions continue *ad infinitum*). Between 100 and 999, there are only thirteen apparent exceptions (for example, 196). I call these *apparent* exceptions only because my smoking PC could not find a palindrome after 100,000 iterations. Perhaps the very next iteration would have yielded one!

The numbers between 10 and 99 yield some further interesting results. Whatever your starting number from this set, the first palindrome you encounter will always be divisible by 11. (Start with 59, for instance, and you will, after three iterations, end up with 1111, obviously divisible by 11; 60 yields 66, 61 yields 77, and so on).

Furthermore, the palindrome you first encounter is always a precise function of the digital sum of the number you began with. (The digital sum of a number is the sum of that number's digits: hence the digital sum of 97 = 16, being 9 + 7 = 16.) If the digital sum of your initial number is, say, 12 (as in the case of 39), the first palindrome you will encounter is always 363. If the digital sum is 17, the first palindrome you encounter is 8,813,200,023,188. Interested readers will be able to fill in the palindromes for the other digital roots.

A feature of reverse addition that appears to hold regardless of the size of the initial number is this: if the initial number is divisible by 11, every subsequent result of reverse-adding is also divisible by 11. Start, for instance, with 396, and the subsequent reverse sums are 1089, 10890, 20691, and so on, all of which are divisible by 11. The same appears to hold if the initial number is divisible by 9 (that is, each subsequent reverse addition yields a result that is divisible by 9) and also if the initial number is divisible by 3.

Digital summing

It is a curious feature of the arithmetic system we use that certain relationships between numbers are preserved when one considers the *digits* that make up those numbers. An example of this parallelism is with numbers and their digital sums and digital roots. The digital sum of a number is the sum of its digits; the digital root is the digital sum reduced, if necessary, to a single digit by repeated addition). For example, the digital sum of 894 is 21 (since 8 + 9 + 4 = 21) and the digital root is 3 (since 2 + 1 = 3).[8]

Consider, firstly, addition. If $a + b = c$, you will find that the digital root of the digital root of a plus the digital root of b equals the digital root of c. Symbolically:

$$D_r c = D_r(D_r a + D_r b).$$

Consider, for instance, 197 + 328 = 525. The digital root (D_r) of 197 is 8 and the D_r of 328 is 4. Hence $D_r a + D_r b$ = 12 and $D_r(12) = 3$. But note, now, that $D_r c$ is also 3 (since 5 + 2 + 5 = 12, and 1 + 2 = 3). This curious identity appears to hold whatever numbers you add together. In other words, the relationship of addition between two numbers is preserved amongst the corresponding digital roots of those numbers.

The case is a little trickier with subtraction. With one exception, the digital root of the result, c, of a subtraction, $a - b$, is equal to the digital root of X where X is the digital sum of a plus 9 minus the digital root of b. Symbolically:

$$D_r c = D_r(D_s a + 9 - D_r b)$$

For example, 673 – 386 = 287. The digital sum of 673 is 16 and the digital root of 386 is 8. Now the digital root of 16 + 9 – 8 is 8, and this is identical to the digital root of the result of the subtraction, 287.

The exception is where b is greater than a and thus the subtraction yields a negative result. Applying our formula above to, say, 613 – 724 = –111 yields: $3 = D_r(10 + 9 - 4) = 6$, which is clearly a contradiction. But a small tweak to our

---

8. Microsoft Excel functions that will quickly calculate the digital sum and digital root of a number are provided in "Calculating digital sums" on page 175.

formula is all that's needed. For subtractions that yield negative answers:

$$D_r c = D_r(D_s a - 9 - D_r b)$$

Let's instantiate that by returning to $613 - 724 = -111$. We now have a true equality: $3 = 3$.

§

On the subject of digital summing, here's a little puzzle:

> Write the numbers 1 to 20 on twenty slips of paper, one number to a slip, and place all the slips in a bowl. Now randomly draw out the slips and place them side by side with the numbers facing upwards. What is the probability that the 31-digit number so formed is divisible by 3?[9]

The solution can be found on page 185.

Magic number 7? Or is it 9?

The number 7 has been considered special throughout history: by scholars and cranks alike. It appears in many guises. There are seven so-called deadly sins, seven wonders of the ancient world, seven wonders of the modern world, seven days of the week, God, apparently, resides above the seventh heaven, and there are seven notes in a musical scale (if you ignore the semitones). Some have even fudged their scientific observations to bring them in line with their numerological superstitions. For example, Isaac Newton could only have observed five colours in the visible spectrum: red, yellow, green, blue

---

9. Taken from Geoffrey Marnell, *Think About It!*, Angus & Robertson, North Ryde, 1989, p. 61.

and violet. But so bewitched was he by the number 7 that he added two more: orange and indigo. Nor is it a bewitchment of the past. In 1956, American psychologist George Miller wrote a paper that went on to become one of the most cited papers in the literature of psychology. It is called "The Magical Number Seven, Plus or Minus Two". In it, Miller summarised the research on the span of human attention. In some of the research Miller cites, subjects were given several stimuli in succession and then asked to immediately recall them. The types of stimuli included binary digits, decimal digits, letters of the alphabet, letters plus decimal digits and monosyllabic words. In one part of the paper, Miller notes:

> "With binary items the span [of immediate memory] is about nine [and it] drops to about five with mono-syllabic English words."[10]

And thus was born the crazy idea that we should never present information to people in more than 7 ± 2. chunks. The idea spread to graphic design, marketing, and even to writing, where the loopy idea was forced on sponge-brained students that there should never be more than 7 sentences in a paragraph, more than 7 paragraphs in a sub-section, more than 7 sub-sections in a section, more than 7 sections in a chapter, and so on.[11] Though well and truly debunked, many still allow their bewitch-ment with the number 7 to cloud their critical faculties. If only they had read the original Miller paper and saw that

---

10. G. A. Miller, "The Magical Number Seven, Plus or Minus Two: Some Limits on our Capacity for Processing Information", *The Psychological Review*, 1956, vol. 63, no. 2, p. 92.
11. The American social scientist Robert Horn commercialised the idea and peddled it under the impressive moniker *information mapping*. See R. E. Horn, *Developing Procedures, Policies & Documentation*, Info-Map, Waltham, 1992.

he was simply averaging the results of many experiments, and thus 7 was merely an arithmetic artefact.[12]

A number due more respect than 7 is 9. It pops up in so many unexpected places in the doodler's mathosphere. As an example, consider an operation I will call *transpose subtraction*. Take any number, shuffle the digits in it in any way you please, and then calculate the difference between the starting number and the shuffled version. The result, it seems, will always be divisible by 9. Table 3 gives some examples.

Table 3: Transpose subtraction

Number	Transposed version	Difference	Multiple of 9
357	753	396	44
735	375	360	40
573	735	162	18
4137	1374	2763	307
15,328	85,123	69795	7755
218,851	588,112	369,261	41,029
5,466,523	3,645,625	1,820,898	202,322

In "Reversal arithmetic" on page 58 I described how you can quickly perform a reverse subtraction on any three-digit number. (A reverse subtraction requires you to reverse the order of the digits and then subtract the smaller

---

12. Moreover, more recent research puts Miller's average closer to 4 than 7, thus leaching some of the mystique from the "magical" number 7. See N. Cowan, "The Magical Number 4 in Short-term Memory: A Reconsideration of Mental Storage Capacity", *Behavioral and Brain Sciences*, 200, vol. 24, p. 87.

of the two from the larger; for instance: 372 – 273.) I also noted that if you continually reverse-subtract using as your starting term each time the result of the previous subtraction, you would always come down to zero.

It is tempting to conjecture that repeated transpose subtraction also leads to zero. Let's start with the case where the starting number has just three digits. If the result of any subsequent transpose subtraction is a two-digit number (as in, for example, 324 – 243 = 81) then it should be clear that there will be no multiple transposing options in any further iteration. There is, to continue our example, only one way to transpose 81, namely as 18. And a little mental arithmetic shows that repeated transpose subtraction of a two-digit numbers eventually leads to a single-digit number—9—which, when subtracted from its transposed self, gives zero. Our conjecture is looking promising, providing only that our sequence of transpose subtractions always yields a two-digit number somewhere in its unfolding.

But further doodling reveals a curious complication. There are least three 3-digit numbers that together can set off an infinite circle of iterations, thus preventing the series from ever reaching a two-digit number, then a one-digit number, and finally zero. These numbers are 459, 495 and 954. The sample set of iterations at the right illustrates this point.

...	...
Absolute difference:	615
Transposition:	156
Absolute difference:	459
Transposition	954
Absolute difference:	495
Transposition:	954
Absolute difference:	459

Continuous loop

Notice that the numbers 459, 495 and 954 are transpositional siblings. They share the same digits. But they have a more interesting property. Each one is capable of creating, with transpose subtraction, a number with exactly the same digits. In other words, if you start with any one of these numbers, it is possible to transpose it in such a way that the absolute difference between the original number and its transposition is a number with exactly the same digits. As the example illustrated on page 66 shows, a transposition of 459 is 954, and the absolute difference between them is 495. Same digits. Further, a transposition of 495 is 954, and the absolute difference between them is 459. Again: same digits. Such an interesting property deserves a name. Let's call the set {459, a 495 and 954} a *self-sufficient triad*.[13]

Our initial conjecture—that repeated transpose subtraction necessarily leads to zero—proved to be incorrect (thanks to a self-sufficient triad). But after a tedious and fruitless slog with pencil and paper, I am willing to propose another conjecture: there is only *one* self-sufficient triad: {459, 495 and 954}.

And now we might consider whether there are self-sufficient quartets, quintets, sextets, septets and so on. Another day, perhaps.

Crabbed addition

Earlier I described how the repeated addition of mirror-image numbers—which I called reverse addition—more often than not leads to a palindrome (that is, a number whose first half is a mirror image of its second half). It is interesting to consider whether the exceptions (such as

---

13. This property is not shared by the other transpositional siblings, namely, 549, 594 and 945, either as a set or in conjunction with any number in the self-sufficient triad.

196) are due to the fact that any carry-forward digit in an addition is carried forward in one-direction only, from right to left, thus limiting the prospect of forming mirror-image numbers. This has *prima facie* plausibility, though further investigation suggests that more could be involved.

If carry-forward digits caused the exceptions, we might expect that reverse additions involving fewer multi-digit additions at each iteration would produce fewer exceptions. To illustrate what I mean, consider the following four varieties of three-digit reverse addition (differing only in the number of places to which the reversed number is moved sidewards before being added to the number from which it is derived):

abc	abc	abc	abc
cba	cba	cba	cba
ghi	jkla	amnba	abccba
Pure	Crab-1	Crab-2	Crab-3

Note that while pure reverse addition involves three multi-digit additions—a + c, b + b, c + a—what I've called *crab-1* reverse addition involves just two—b + c, c + b—*crab-2* one—c + c—and *crab-3* none.[14] What makes this set of additions potentially useful in judging the effect of carry-forwards on reverse-addition palindromes is that each type yields, more often than not, a palindrome after repeated reverse

$$
\begin{array}{r}
174 \\
+\ 471 \\
\hline
1211 \\
+\ 1121 \\
\hline
13231 \\
\end{array}
$$

Crab-1 example

------

14. We are ignoring here the potential carry-forward addition *a* + 1 to yield *j* with crab-1.

addition (with crab-3 reverse addition obviously yielding a palindrome in all cases).

Since the likelihood of there being a carry-forward digit in an addition decreases as the number of multi-digit additions decreases, we might expect that the number of non-palindrome exceptions decreases as we move from pure reverse addition through crab-1, crab-2 to crab-3 reverse addition. (The fact that crab-3 yields no exceptions encourages this suspicion.)

However, this is not the case. While there are 13 exceptions in pure reverse addition for the numbers 1 to 1,000, for that same range crab-1 yields 22 exceptions and crab-2 208 exceptions. In other words, fewer carry-forwards yield—in pure, crab-1 and crab-2 reverse addition, at least—a greater number of exceptions, a seemingly counter-intuitive result. Unidirectional carry-forward may not, then, be the cause of the non-palindromes (although, of course, we have not strictly compared likes with likes).

What does decrease as the number of multi-digit additions decreases is the maximum number of iterations needed to reach a palindrome: in pure reverse addition, no more than 24 iterations are needed to produce a palindrome, dropping to 11 for crab-1, 8 for crab-2 and 1 for crab-3.

Interestingly, many palindromes are repeated in pure and crab-1 reverse addition (while each palindrome is unique in crab-2 and crab-3 reverse addition). In at least some cases you can deduce how many times a palindrome will be repeated from the digital sum of the starting number.

Fun with a tree of differences

Here is a fairly simple mathematical trick. Ask someone to think of a three-digit number in which no digit is repeated. (376 meets this requirement, but 377 does not, as there are two sevens.) Get them to write the number down out of your sight and to derive all five three-digit permutations of the number they thought of. (To continue our example, the three-digit permutations of 376 are 367, 736, 763, 637 and 673.) Next ask your companion to write all six numbers in a row starting from the smallest number, continuing with the next smallest number and so on, ending with the largest number, as in:

$$367 \quad 376 \quad 637 \quad 673 \quad 736 \quad 763$$

Now ask your companion to subtract adjacent pairs of numbers and to keep doing so until they come down to just one number, as in:

367		376		637		673		736		763
	9		261		36		63		27	
		252		−225		27		−36		
			−477		252		−63			
				729		−315				
					−1044					

Once your companion has reached the solitary figure at the bottom of subtraction triangle — which I shall call the *concluding term* — ask them to tell you the three-digit number they originally thought of. In a second or two of mental arithmetic (or rote recall) you tell them the concluding term they arrived at. For example, they say 367, you immediately say −1044 (as in our example); or they say 825 and, in a flash, you say −1566.

So how is it done? The key to quickly calculating the concluding term is the difference between the largest digit in the starting number and the smallest digit, which for brevity's sake I shall call the *limdiff*. In our worked example, the limdiff of 367 is 7 – 3 = 4.

Now for the secret to this trick: the concluding term is *always* the limdiff multiplied by –261. In our example, the limdiff is 4, so the concluding term is 4 × –261 = –1044.

Calculating the limdiff is pretty straightforward, but for those who shy away from multiplication, the following table will help if you prefer to memorise the concluding term for each of the 8 limdiffs that cover all the unique-digit–three-digit numbers.

Limdiff	Concluding term	Limdiff	Concluding term
2	–522	6	–1566
3	–783	7	–1827
4	–1044	8	–2088
5	–1305	9	–2349

One final example: your companion says 581. You quickly calculate the limdiff as 7 (being 8 – 1) and then multiply –261 by 7 to get –1827 (or merely recall –1827 from the memorised table).

Incidentally, the digital sum of every number in a subtraction triangle formed from a permuted three-digit number is nine (irrespective of whether the permutations are ranked).

Warning: unless you have plenty of time on your hands, I don't recommend exploring this topic further with four-digit numbers. While three-digit numbers give

you a starting row of just 6 ranked numbers, four-digit
numbers require a starting row of 24 ranked numbers.
And the subsequent calculations get tedious, with a
typically enormous concluding term. For example, 1234
yields –614,015, 010 and 2713 yields –594,147,465. Perhaps
there is an opal in the rock, but now we are well and truly
outside the realm of doodling.

Repeated alternating reverse addition and reverse subtraction

If reverse addition—that is, the addition of mirror-image
numbers—more often than not leads to a palindrome, and
reverse subtraction always leads to zero, what might be
the result if we repeatedly alternated reverse addition and
reverse subtraction? At the start, let's give this process a
name—RARARS: *Repeated Alternating Reverse Addition and
Reverse Subtraction*.

RARARS yields a curious type of number. I shall call
it a *circular number* (and explain why in just a moment).
Further, *every* RARARS process produces a circular
number, and yet there are comparatively few circular
numbers. For instance, only five circular numbers appear
in the RARARS iterations of the first 10,000 positive
integers. In other words, if you start with any number
from 1 to 10,000, you will always find, in your string of
RARARS iterations, one of just five circular numbers.
These numbers are 0, 1089, 10,890, 10,989 and 109,989.

Suppose, for example, that you start with 1792. Its
reversal is 2971. Adding these two numbers yields 4763
which, when reversed, becomes 3674. Taking the smaller
from the larger of these last two numbers yields 1089, one
of the circular numbers.

This number—1089—appears in RARARS iterations more often than any other circular number. It also appears with uncanny frequency in the RARARS iterations of numbers greater than 10,000.

The reason I've called these numbers *circular* numbers is because they are the first number in an infinite circle of repeating numbers. Take 1089, for example. Assume, firstly, that the next operation is reverse addition. The result is 10,890 (that is, 1089 + 9801). The next RARARS operation is reverse subtraction, which yields 1089 (being 10,890 – 9801). The next RARARS operation

Starting number:	1792
Reverse:	2971
Addition:	4763
Reverse:	3674
Absolute difference: 1089 ←	
Reverse:	9801
Addition:	10,990
Reverse:	9901
Absolute difference: 1089 ─	

Continuous loop

obviously just repeats what has gone before. Thus we have an infinitely repeating circle of numbers: 1089, 10,890, 1089, 10,890 and so on.

If the next operation after first reaching 1089 had been reverse subtraction rather than reverse addition, the next number would be 8712 (that is, 8901 – 1089), the next would be 10,890 (8712 + 2178), the next would be 1089 (10,980 – 9801), the next is 10,890 (1089 + 9801), the next is 1089 (10,890 – 9801), and so on. In other words, 1089 followed by a reverse subtraction yields 1089 followed by a reverse addition which, as we've seen, forms a repeating circle of numbers.

Similarly, 10,890 forms an infinite circle with 1089; 10,989 forms a repeating circle with 109,890; 109,989 forms an infinite circle with 1,099,890; and, of course, 0 forms an infinite circle with itself.

On the basis of an examination of the first 10,000 positive integers, induction leads me to conjecture that all positive integers yield a circle of infinitely repeating numbers when subjected to a RARARS process.

Zeroing in, yet again

I have noted the affinity subtraction has with zero. Put subtraction to work in one way or another and, more often than not, zero will pop up in your calculations. Here is another example of this peculiar affinity.

Take any two numbers (say, A and B). Subtract the lesser from the greater of the two (yielding C). Now consider B and C and, again, take the lesser from the greater (yielding, let us say, D). Do the same with C and D (yielding E), then with D and E and continue in like fashion. You will, I suggest, always come to zero. In fact, you will always come to a repeating triplet of numbers comprising zero and two instances of the last number you encountered before reaching zero.

For example, take 312 and 520 as our random starting numbers. The first number our operation gives is 208 (being 520 – 312), followed by 312 (520 – 208), followed by 104 (312 – 208), followed by 104 (208 – 104), followed by zero. Continuing in like fashion produces the repeating triplet {0, 104, 104 ...}.

	312
[abs –]	520
	208
[abs –]	312
	104
[abs –]	208
	104
[abs –]	104
	0

Iterative subtraction of this sort displays three curious features. Firstly, it does not matter whether you start with A and B or B and A. (In our example, you could have subtracted 208 from 520 or from 312 at the start.) The numbers in the chain of subtractions will not be identical, but you will still encounter zero and the same repeating non-zero number. To show this, let A = 520 and B = 312. The iterations this time yield 208, 104, 104 and 0.

Secondly, *every* result in the chain of subtractions is perfectly divisible by the non-zero repeating number. (In our example, 104 divides perfectly into 312, 208 and 104.)

Thirdly—and most interesting of all—the non-zero repeating number appears always to be the greatest number that will divide perfectly into *both* your starting numbers. (Note that 104 is the greatest number that can divide perfectly into 520 and 312.) In other words, the non-zero repeating number is the *highest common factor* (or *greatest common divisor*, as it is also known).

As a further illustration of these features of iterative subtraction, take the numbers 806 and 624. Iterative subtraction yields the set of numbers {182, 442, 260, 182, 78, 104, 26, 78, 52, 26, 26, 0 ...}. Reversing the two starting numbers still gives you 26 as the repeating non-zero number (although the number of iterations needed to reach this number is now 14 rather than 12). Furthermore, every number in the set of intermediate subtractions is divisible by the non-zero repeating number. (That is, all these numbers are perfect multiples of 26.) Finally, a spot of mental arithmetic—or a good calculator—will show that no number greater than 26 will divide perfectly into both 806 and 624.

From limdiff to diffdiff

Earlier I wrote of a trick whereby you get a companion to write down a three-digit number with no repeating digits, derive all five three-digit permutations of the number, rank all six numbers starting from the smallest number and ending with the largest, and then repeatedly subtract adjacent pairs of numbers until they come down to just one number. I noted that on being told the original number, you can very quickly calculate the final number your companion will have arrived at. The key to the trick was to calculate what I called the *limdiff* of your companion's starting number and multiply it by –261. The limdiff is the largest digit in the starting number minus the smallest, so the limdiff of, say, 571 is 6 (being 7 – 1) and the limdiff of 367 is 4 (being 7 – 3).

There are a number of variations on this trick. If instead of asking your companion to repeatedly subtract adjacent pairs of numbers you ask them to subtract adjacent numbers in the first row of numbers, add adjacent pairs in the next, subtract adjacent pairs in the third, add in the fourth and subtract the final pair to reach a concluding term, the number they will have arrived at is always the limdiff multiplied by –9. Here is an example:

367		376		637		673		736		763
	9		261		36		63		27	
		270		297		99		90		
			27		–198		–9			
				–171		–207				
					–36					

Note that the limdiff is 4 and 4 × –9 = –36, the concluding term in the triangle of numbers. The simplicity of the calculation ensures that this trick can be mastered by pretty well anyone.

What makes this especially simple is that, for three-digit starting numbers, there can only be 8 limdiffs and they are all single digits: 1 through to 8.

Note too that the digital sum of every number in the results (that is, from row 2 onwards) is 9.

A slightly more complex variation on the theme is to change the order of the arithmetical operations: ask your companion to add first, rather than subtract, and to alternate the operations. That is, instead of performing the set of operations {−, +, −, + and −} they perform the set {+, −, +, − and +}. An example follows, using 629 as the starting term needing to be ranked:

269		296		629		692		926		962
	565		925		1321		1618		1888	
		360		396		297		270		
			756		693		567			
				−63		−126				
					−189					

The key to being able to quickly calculate the concluding term when told the starting term is a number I'll call the *diffdiff*. The diffdiff is the difference between the differences between the ranked digits in the starting term. The ranked digits of 629 are 2, 6 and 9. The diffdiff is $(9 − 6) − (6 − 2) = −1$. The trick is complete when you multiply the diffdiff by 189. And −189 is indeed the concluding term in our inverted triangle of calculations above.

By the way, it makes no difference to the diffdiff whether you rank the digits in ascending order (as above) or descending order. If, for our starting number of 629, we instead ranked the digits 9, 6 and 2, the diffdiff is still −1, since $(9 − 6) − (6 − 2) = −1$. Indeed, the calculation is easier if you rank the digits in descending order.

Finally, note that every number in the subtraction triangle from the second row onwards is divisible by 9.

A final example: your companion says 291. You quickly calculate the diffdiff as $(9 − 2) − (2 − 1) = 6$ and then multiply 189 by 6 to get 1,134. And here's the triangle of calculations to prove that the answer is correct.

129	192	219	291	912	921
321	411	510	1203	1833	
90	99	693	630		
189	792	1323			
603	531				
1134					

Paired digital sums

A curious feature of any number is that the difference between it and the sum of its digits is always divisible by 9. For example, take 37,213 and add its digits together. The result is 16. Now 37,213 minus 16 is 37,197 which is perfectly divisible by 9. The result is the same whether you subtract the digital sum (as in our example) or the digital root (being the digital sum reduced to one digit by iterative digital addition). To continue our example, the digital root of 37,213 is 7 (being the sum of the digits in the digital sum) and 37,216 is also perfectly divisible by 9.

This feature of numbers provides a further way of showing the affinity between subtraction and zero. For if you take a number — any number — and subtract that number's digital sum (or digital root) and keep doing so using the result of your previous subtraction each time, you will always come to zero. Moreover, each intermediate result is perfectly divisible by 9 and the digital sum of each intermediate number is divisible by 9. For example, start with 128 and subtract the digital sum (11). The sequence of numbers generated by repeating the process is {117, 108, 99, 81, 72, 63, 54, 45, 36, 27, 18, 9, 0} each of which is divisible by 9.

Now consider a special form of digital sum I'll call *paired digital sum* (or PDS). The PDS of a three-digit number abc is $[a + b, b + c]$. For example, the PDS of 341 is 75 (being 3 + 4 conjoined with 4 + 1). If the second addition ($b + c$) results in a number greater than 9, the tens digit is added to the first addition (much as digits are carried forward in ordinary addition). Thus the PDS of 539 is 92 and the PDS of 879 is 166. Similarly, the PDS of a four-digit number $abcd$ is $[a + b, b + c, c + d]$, again with a carryover added where necessary. So the PDS of 1234 is 357 and the PDS of 8929 is 1821. The procedure is similar for greater numbers: PDS(62,178) = 8,395, PDS (97,625) = 17,387, and so on. Obviously, the PDS of a 1-digit or 2-digit number is the same as the number's digital sum.

As with iterative subtraction of common digital sums, iterative subtraction of PDSs also forms a sequence ending in zero. Take, for example, 689. PDS(689) = 157 and 689 − 157 is 532. Further, PDS(532) = 85 and 532 − 85 = 447; PDS(447) = 91 and 447 − 91 = 356; PDS(356) = 91 and 356 − 91 = 265; PDS(265) = 91 and 265 − 91 = 174; PDS(174) = 91 and 174 − 91 = 83; PDS(83) = 11 and 83 − 11 = 72; PDS(72) = 9 and 72 − 9 = 63; PDS(63) = 9 and 63 − 9 = 54 ... after which it should be obvious that the sequence peters out with consecutive diminishing

	689
[-PDS]	157
	532
[-PDS]	85
	447
[-PDS]	91
	356
[-PDS]	91
	356
	...
	0

multiples of 9 leading finally to zero. As is the case in this example, 9 and 91 appear with uncanny frequency in the iterative subtraction chains of a great many numbers. For four-digit numbers, 827, 1736, 2645 and 3554 appear frequently, as do 91 and 9.

For three-digit numbers, the minimum number of iterations required to reach 0 is 10. In fact, the number of iterations depends on which hundreds grouping the starting number belongs to. Between 100 and 199, the number of iterations is 10; between 200 and 299 it is 11; between 300 and 399 it is 12; between 400 and 499 it is 13; and likewise up to the 900–999 grouping, which required 18 iterations to reach 0. Note the pattern?

Corresponding digital addition

Take any two three-digit numbers. Let's represent them as *abc* and *def*. Now add corresponding digits, that is, $a + d$, $b + e$ and $c + f$, reducing the result each time to its digital root. (The digital root of a number is the sum of its digits reduced, if necessary, to a single figure by further addition. Thus the digital root of 3 is 3 and the digital root of 17 is 8.) If our starting numbers are 362 and 739, our results are $3 + 7 = 10$, $1 + 0 = 1$; $6 + 3 = 9$ and $2 + 9 = 11$, $1 + 1 = 2 = 2$. Concatenate these three results—that is, string them together—to form a new three digit number: 192. Let's represent this third number as *ghi*.

Now repeat the process using the first number you jotted down and the number you have just derived—that is, using *abc* and *ghi*—and continue doing so until you have derived nine three-digit numbers.

Finally, add up all the first digits in the series of nine derived numbers, all of the second digits in that series, and then all of the third digits. You will then have three two-digit totals—say, x, y and z. Curiously, x, y and z will always be a multiple of 9—except when b or c is 0, in which case y or z will be 0.

Suppose, for instance, you select 362 and 739 as starting numbers and proceed to derive, by repeated corresponding digital addition (CDA), the series {192, 464, 736, 198, 461, 733, 195, 467 and 739}. You will find that the sum of all the first digits is 36, of all the second digits, 54, and of all the third digits, 45. Note that each sum is a multiple of 9.

3	6	2
7	3	9
10	9	11
+		+
1	9	2
4	15	4
+		
4	6	4
	. . .	

There are many curious features of CDA. For a start, the derived numbers always repeat. If the first of the two starting numbers is 999, then each derived number is, obviously, a repetition of the second starting number. If the first starting number is not 999 yet *every* digit in that number is divisible by 3 (as in 699 and 369), the derived numbers repeat after every *third* addition. (For example, the starting numbers 369 and 531 yield 891, 261, 531, 891, 261, 531 and so on). In every other case, the derived numbers repeat after every *ninth* addition. This is the case in the example above where the starting numbers are 362 and 739.

Another curious feature of CDA has to do with the digits that appear in the derived numbers. If the *n*th digit in the first of the two starting numbers is 3 or 6, the *n*th digits in each of the derived numbers will always be from one of only three sets of repeating triplets — {1, 4, 7}, {2, 5, 8} and {3, 6, 9} — the digits of which can be in any order. Consider again the example given above, where we start with 362 and 739. Note that the first digit in the first number is 3 and the first digits in the derived numbers form the repeating triplet {1, 4, 7}. Note too that the second

3	6	2
7	3	9
1	9	2
4	6	4
7	3	6
1	9	8
4	6	1
7	3	3
	. . .	

digit in the first number is 6 and the second digits in the derived numbers form the repeating triplet {9, 6, 3}.

Further, if the nth digit in the first of the two starting numbers is 9, the nth digits in each of the derived numbers will all equal the nth digit of the second starting number. But especially curious is the fact that if the nth digit in the first of the two starting numbers is not 0, 3, 6 or 9, the nth digits in the derived numbers form repeating sets of *all* nine digits from 1 to 9 (in some order or other). Take, for instance, 569 and 318. The first digit in the first starting number is not 0, 3, 6 or 9 and the first digits of the derived numbers form the repeating set {8, 4, 9, 5, 1, 6, 2, 7 and 3}.

We have here the ingredients for a party trick. Ask as colleague to jot down two three-digit numbers (and, to make your task easier, ask them to pick numbers without a zero in them). Then get them to add corresponding digits, reducing the result each time to its digital root. Get them to keep doing likewise, using their first number as the number to add to in each case, and to stop after nine additions. (In other words, just do what we have been doing throughout this section.)

Now get your colleague to add up all the first digits in the nine derived numbers, all the second digits, and all the third digits. We noted earlier that these sums—which we labelled x, y and z—will be multiples of 9, but the trick is this: when your colleague tells you their two starting numbers—abc and def—you can immediately tell them what x, y and z is. This is how:

- If $a = 9$, $x = 9d$.
- If $b = 9$ and $e = 9$, $y = 36$; otherwise $y = 9e$.
- If $c = 9$ and $f = 9$, $z = 36$; otherwise $z = 9f$.
- If a, b or $c = 3$ or 6, calculate the digital root of it and the corresponding digit in the second number (d, e

or *f*). If the digital root is:
- 1, 4 or 7, the corresponding sum (*x*, *y* or *z*) is 36
- 2, 5 or 8, the corresponding sum (*x*, *y* or *z*) is 45 or
- 3, 6 or 9, the corresponding sum (*x*, *y* or *z*) is 54
- in all other cases, *x*, *y* and *z* are 45.

For example, your colleague says 571 and 872 and, noting immediately, that there is no 3, 6 or 9 in the first number, you reply in a flash that the three sums are 45, 45 and 45. If the numbers had been 613 and 894, you calculate 6 + 8 = 14, 1 + 4 = 5 (which belongs to that repeating triplet that sums to 45), therefore *x* = 45. Since *b* is neither 3, 6 or 9, *y* must be 45. For *c*, calculate 3 + 4 = 7, therefore *z* = 36 (referring again to our repeating triplets). Finally, consider 493 and 679. *x* must be 45, *y* = 9 × 7 = 63 and *z* = 54 (since 3 + 9 = 12, 1 + 2 = 3 and belongs to that repeating triplet that sums to 54).

Corresponding absolute digital subtraction

Take any two numbers with the same number of digits not one of which is zero. The numbers 362 and 788 meet this criterion and, for the sake of illustrations, I'll represent these number as *abc* and *def*.

Now perform what I'll call *corresponding absolute digital subtraction* (CADS, for short); that is, take each pair of corresponding digits — {*a*, *d*}, {*b*, *e*} and {*c*, *f*} — and subtract one digit from the other (ignoring the sign in the case of a negative result). Suppose that the results are *g*, *h* and *i* respectively. Now form a new number by concatenating the three results. The new number in our example will be *ghi*. Now repeat the process again, taking as your two terms the first of your starting numbers — *abc* — and

the new number you have derived: *ghi*. Keep doing
likewise. You will quickly come to a pair of repeating
numbers whose sum is the first of the two numbers you
started with (the one from which you have been
repeatedly subtracting the results of your absolute digital
subtractions).

   Take, for example, 362 and 788. CADS
yields 426. Now do the same with 362 and 426.
The result is 144. Repeating the process yields
the repeating pair {222, 140}, the sum of which
is 362, the number you started with.

   Two more examples: 36 and 92 yield 64, 32,
4, 32, 4, 32 and so on. The two repeating
numbers — 32 and 4 — add up to the first
number: 36. Further, 1263 and 3452 yield 1052,
211, 1052, 211 and so on. Note that 1052 plus
211 yields the starting number: 1263.

3	6	2
7	8	8
4	2	6
1	4	4
2	2	2
1	4	0
2	2	2
1	4	0
.	.	.

3	6	2
7	8	8
4	2	6
3	6	2
4	2	6
3	6	2
4	2	6
3	6	2
.	.	.

You get a similar result if, instead of repeatedly
subtracting the results of your absolute digital
subtractions from the first number you started
with, you subtract them from the second
number. In our first example — 362 and
788 — the results are 426, 362, 426, 362 and so
on. Notice that the sum of the repeating pairs is
now equal to the second of the two starting
numbers: 788.

A further curious feature of CADS can be seen
if one looks at corresponding numbers in the
sequence of numbers generated when one subtracts from
the first starting number and when one subtracts from the
second starting number. With very few exceptions, the
sum of the absolute differences between corresponding
repeating numbers is equal to the absolute difference
between the two starting numbers.[15] In the example
illustrated above, the two repeating pairs are {222, 140}

and {426, 362}. The absolute difference between corresponding numbers in these pairs yields {204 and 222}. The sum of these two numbers is 426, which just happens to be the absolute difference between the two starting numbers: 362 and 788.

Repeated digital multiplication

Take any three-digit number composed of the digits 1 to 9 (say, *abc*). Now multiply each of the digits together in this order: $a \times b$, $a \times c$ and $b \times c$). Where the result of a multiplication is greater than 9, add the digits of the result together (and do likewise if the result of the addition is greater than 9). Now form a new three-digit number by concatenating the three results so far obtained in the order in which they were obtained. For instance, if your starting number is 374, you will have $3 \times 7 = 21$, $2 + 1 = 3$; $3 \times 4 = 12$, $1 + 2 = 3$ and $7 \times 4 = 28$, $2 + 8 = 10 = 1 + 0 = 1$. Concatenating the three results gives you 331.

Now do to your new number what you did to your starting number. The result is 933 (that is, $3 \times 3 = 9$; $3 \times 1 = 3$ and $3 \times 1 = 3$). Do likewise to your new number and keep doing so until you encounter the following curious feature, namely, one result, or a subset of results, will begin to repeat. To continue our example, after 993 you get 999, 999, 999 and so on *ad infinitum*.

Obviously, if you start with 999 your repetition begins immediately. Likewise 333 and 666. For other starting numbers, the

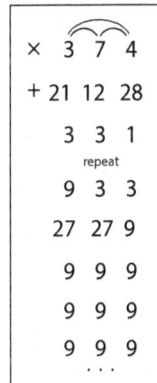

×	3	⌢7	4⌢
+	21	12	28
	3	3	1
		repeat	
	9	3	3
	27	27	9
	9	9	9
	9	9	9
	9	9	9
		. . .	

---

15. The exceptions seem to appear when either number has a zero for a digit.

repetition can begin as early as the second iteration or as late as the seventh. For example, 125 yields 251 which in turn yields 125 and so on. Similarly, 741 yields 174 which yields 741 etc. In some cases, the repeating subset begins with the number you started with. Some examples are 287, 752, 851, 485, 524, 128, 287, 752. (In these cases, the repetition begins after the sixth multiplication. with other numbers the subset begins with the first result, as in 721, 572, 815, 845, 542, 218, 278, 572, 815. (Again, the repetition with the sixth repetition.)

Curiously, some repetitions occur immediately and repeat the starting number. Some examples: 111, 215, 512 and 999. In fact, a pencil, the back of a coaster and a few idle moments is all it will take show that no other three-digit number generates itself on the first iteration of repeated digital multiplication. We noted earlier some special three-digit numbers we christened *self-sufficient triads* (see page 67). Perhaps 111, 215, 512 and 999 too deserve a special name, say, *self-replicating triads*.

There are other curious features of repeated digital multiplication. For instance, if your starting number has a digit that is divisible by 3 — it does not matter if it is the first, second or third digit — and no zeros, then your repeating set is always composed solely of the number 999. (Conversely, if you reach 999, you will have started with a number containing at least one digit divisible by 3 and no zeros.)

Further, if a starting number abc results in a repeating subset starting with jkl, then the starting number composed of the digits $9 - a$, $9 - b$ and $9 - c$ will also produce a repeating subset starting with jkl. For example, 154 yields a repeating subset beginning with 542; and so does 845. Furthermore, no other starting number will produce a repeating subset beginning with 542.

Repeated digital addition

In the last section I considered some of the curious features of the repeated digital multiplication of three-digit numbers. In this section I'll consider some curious features of a companion operation: repeated digital addition.

The digital addition of a three-digit number abc is def where d is the digital root of $a + b$, e is the digital root of $a + c$ and f is the digital root of $b + c$. Repeated digital addition is the application of digital addition to the result of each digital addition *ad infinitum*. Note that unlike corresponding digital addition — considered on page 80 — repeated digital addition begins with one, not two, three-digit numbers.

As with repeated digital multiplication, repeated digital addition (RDA) of any three-digit number results in a circle of repeating numbers. For example, RDA(159) yields {615, 726, 948, 483, 372, 159, 615 ...}. Note the repetition from the fifth term of the set of results. Some numbers produce a repetition at the second step. For example, RDA(144) = {558, 144, 558, 144 ...}.

+	1	5	9
6	1	5	
7	2	6	
9	4	8	
4	8	3	
1	5	9	
6	1	5	
. . .			

Continuous loop

Where the repetition begins depends in part on the digital root of the starting number: $D_r(n)$. If $D_r(n) = 9$, repetition begins at the first or second iteration. For example, RDA(513) = {684, 513 ... If $D_r(n) = 3$ or 6, repetition begins at the sixth or seventh iteration. In all other cases, repetition begins at the eighteenth or nineteenth iteration.

A little mental arithmetic will show that if $D_r(abc) = 9$ and $b = 9$, repetition begins immediately. For example, RDA(693) yields {693, 693 ...}.

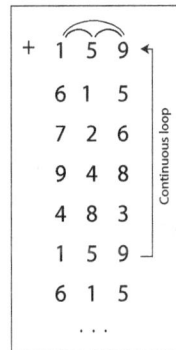

Another feature of RDA is that if $D_r(n) = 9$, the digital root of each number in the circle of repeating numbers is also 9.

Furthermore, the number of three-digit numbers that show repetition after one or two iterations is 100; the number that show repetition after six or seven iterations is 200 and the number that show repetition after eighteen or nineteen iterations is 600. Curiously, in all instances except those that repeat after 7 or 19 iterations, it is the starting number that begins the circle of repeating numbers. In those that repeat after 7 or 19 iterations, the number immediately following the starting number begins the circle of repeating numbers.

Another curious feature is that in the majority of cases—the 593 cases where repetition occurs after 18 iterations—each number in the set {1, 2, 3, 4, 5, 6, 7, 8, 9} is repeated once in the set of first-digit numbers, once in the set of second-digit numbers; and once in the set of third-digit numbers of the numbers in the repeating circle.

Repeated paired-digit addition

Here is a variation on corresponding digital addition, discussed on page 80. Take any two three-digit numbers (*abc* and *def*) and calculate the digital root of each corresponding pair of digits, that is, $D_r(a + d)$, $D_r(b + e)$ and $D_r(c + f)$. Suppose the individual results are *g*, *h* and *i*. Concatenate these digits into the number *ghi*. Repeat the process using *def* and *ghi* (yielding *jkl*) and again (using *ghi* and *jkl*).

The diagram at the top of page 89 illustrates the process, using 462 and 179 as the two starting numbers (that is, as *abc* and *def*.

$$abc \quad def = ghi = jkl = mno = \dots$$

462   179   542   622   264   ···

If you keep doing likewise, you will find—with one exception—that your developing chain of numbers, starting with your two initial numbers (*abc* and *def*), repeats from step 24 onwards.

The one exception is where both starting numbers are composed of digits each of which is divisible by 3 (as in the case of 336 and 669). In these cases, your two starting numbers begin to repeat from step 8 onwards. (For example, 336 and 669 yield 996, 666, 663, 339, 993, 333, 336, 669 and so on.)

There are many curious features of repeated paired-digit addition. For instance, if the first of the two starting numbers is composed of digits each of which is divisible by 3 (as in the case of 963 and 623), the numbers derived at steps 1, 9 and 17 are the same (686 in this particular example). So too are the numbers derived at steps 5, 13 and 21 (313 in our example). Moreover, the sum of these two repeating numbers is always 999 (686 + 313). Furthermore, the number at every fourth step beginning with step 3 is composed of digits each of which is divisible by 3.

The case is similar if the second, but not the first, of the starting numbers is composed of digits each of which is divisible by 3 (as in 614 and 693). The numbers derived at steps 2, 10 and 18 are the same (911 in this example) and the numbers derived at steps 6, 14 and 22 are the same (in this case, 988). Again, note that the two repeating numbers add to 999. Similarly, the number at every fourth step, this time starting from step 4, is composed of digits each of which is divisible by 3.

An apparent exception to the observation that all three-digit numbers repeat after step 24 occurs when one or both starting numbers contains a zero. For instance, 105 and 345 yield 195 and 345 at steps 24 and 25. But in digital addition, 0 and 9 are identical: $3 + 9 = 3$ and $3 + 0 = 3$. Thus 105 and 195 can be considered interchangeable and the hypothesis of repetition after step 24 is met.

Alternating digital operations

In the last few sections I have been exploring the curious properties of repeated digital subtraction, repeated digital multiplication and repeated digital addition. As we might expect, properties no less curious can be found if we combine a number of digital operations into the one repeated iteration.

For example, take any three-digit number where the digits are not all the same. (219 and 229 meet this qualification; 222 does not.) Now perform digital addition on your number followed by digital subtraction on the result of your digital addition, followed by digital addition on the result of your digital subtraction, and so on in like fashion, alternating between digital addition and digital subtraction. (The digital addition of a number abc is def where $d = a + b$, $e = a + c$ and $f = b + c$, with any double-digit result reduced to a single digit by repeated digital summing. The digital subtraction of abc is jkl where $j = a - b$, $k = a - c$ and $l = b - c$, with the sign ignored in each case.)

Consider, for example, the number
219. Repeated alternating digital addition
and subtraction yields the set of numbers
{321, 121, 323, 101, 121, 101, 121 ...}. Note
that 101 sets off a circle of repeating
numbers. Indeed, you will find that any
three-digit number—subject only to the
qualification that the digits are not all
identical—will lead to a repeating circle of
numbers.

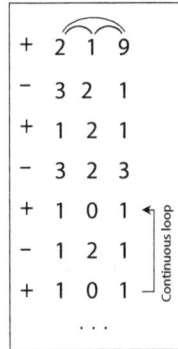

+	2	1	9
−	3	2	1
+	1	2	1
−	3	2	3
+	1	0	1
−	1	2	1
+	1	0	1
		...	

Continuous loop

What is especially curious about these
repeating circles is that the number that starts off the
circle in the chain of iterations is invariably composed of a
zero and another digit repeated. For example, in the
instance considered above, the circle begins with 101. To
take another example, 825 yields the set of numbers {147,
363, 969, 303, 363, 303, 363 ...} with the repeating circle
kicking off with 303: a number composed of a zero and a
repeating digit. Similarly, 185 and 376 yield repeating
circles that kick off with 110 and 440 respectively.

It does not matter whether you start with digital
subtraction or digital addition. You still come to a
repeating circle of numbers kicked off by a number
composed of a zero and a repeating digit. In this case, the
number that kicks off the circle is different, although
curiously it is composed of the same digits. For example,
144 yields 033 as the kick-off number if digital addition
preceded digital subtraction, and 330 if digital subtraction
preceded digital addition.

You will also encounter circles of repeating numbers
if you combine other types of digital arithmetic. For
example, if you apply alternating digital multiplication
and digital addition to 825, the result is {741, 285, 714, 825,
741, 285 ...}. More complex combinations of digital
arithmetic also yield repeating circles. For example,

applying alternating digital subtraction followed by digital addition followed by digital multiplication to 278 yields the set {561, 267, 356, 231, 534, 623, 431, 754, 832, 561, 267, 356 …}.

Clearly there is much more that we could explore here but, sadly, *tempus fugit*.

Cross-adding

Take any two numbers (say 74 and 62). Set out the numbers in column format, as though you wanted to add them:

74

62

Now imagine diagonal lines cutting through each number, one set of lines going from right to left, another left to right.

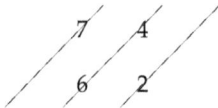   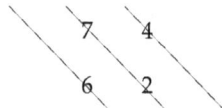

Set 1   Set 2

In the first set, the lines cut through 7, 4 and 6, and 2 (working from top-right to bottom-left). Where the line cuts more than one digit, add the digits together (leaving us with 7, 10 and 2 in our example); and where any sum is more than one digit long, carry the greater digit over to the adjoining digit as you would do in normal addition. Thus 7, 10 and 2 becomes 8, 0 and 2.

Now do the same with the other set of diagonals, those running downwards from left to right. This yields 6, 9 and 4.

If we treat our two results as numbers in their own right rather than as sets of unrelated digits, we have the numbers 802 and 694. Note that the absolute difference between these two numbers — 108 — is exactly 9 times the absolute difference between the two starting terms:

$$9 \times (74 - 62) = 108$$

Now continue the same cross-add operations using the two numbers you derive at each step.[16] The derived numbers, and their differences, are:

74		
62	=	12
802		
694	=	108
8714		
7742	=	972
94,882		
86,134	=	8,748
...		...

In this example, the absolute difference between the starting numbers and between each set of numbers forms the series {12, 108, 972, 8,748 ...}. Let's call this the *sequence of differences*. Note that in each case the absolute difference

---

16. To ensure that every digit in a number is crossed by a diagonal, there needs to be (n + 1) diagonals per set for an n-digit number. Thus 802 and 694 will need four diagonals per set. The first set will cross 8, 0 and 6, 2 and 9, and 4. The second will cross 6, 8 and 9, 0 and 4, and 2.

between the *derived* pairs is 9 times the absolute difference between the two numbers that, when cross-added, formed that pair. For example, take as our starting numbers 802 and 694 (which differ by 108). Right-to-left diagonal adding followed by left-to-right diagonal adding yields the numbers 8714 and 7742. These two numbers differ by 972. Notice that 972 is exactly nine times 108.

The differences between the pairs of numbers generated by cross-adding can be expressed as:

$$9^{n-1} \times \text{ABS}(a - b)$$

where n is the position of any particular difference in the sequence of differences and $\text{ABS}(a - b)$ is the absolute difference between the two numbers that were first cross-added (being 74 and 62 in our example). Thus the fourth number in the sequence of differences generated by 74 and 62 is $9^3 \times 12 = 8748$, the very number we got above by manual cross-adding.

Curiously, regardless of the two numbers you start with, the relationship between them and the derived number is the same. For another example, consider 26 and 83. Manually cross-adding yields the following sequence:

26		
83	=	57
343		
856	=	513
4286		
8903	=	4617
51,763		
93,316	=	41,553
...		...

Notice again that the absolute difference between any pair of derived cross-additions is 9 times the absolute difference between the two numbers that formed them. And as before, the differences between the pairs of numbers generated by cross-adding can be expressed as:

$$9^{n-1} \times \text{ABS}(a - b)$$

Thus the third number in the sequence of differences {57, 513, 4617, 41,553 ...} is $9^3 \times 57 = 41,533$.

After much pen-and-paper drudgery, I'm tempted to conjecture that cross-adding x with y *always* generates a pair of numbers that differ by 9 times the difference between x and y (signs ignored).

§

Let's modify our cross-adding explorations and ponder what the result might be if the starting and derived numbers are *added* to each other rather than subtracted. A spot of doodling shows that the sum of the two numbers in a derived pair appears always to be 11 times the sum of the numbers that generated that pair by cross-adding. For instance:

$$
\begin{array}{rcl}
74 & & \\
62 & = & 136 \\
\\
802 & & \\
694 & = & 1496 \\
\\
8714 & & \\
7742 & = & 16{,}456 \\
\\
\cdots & & \cdots
\end{array}
$$

Moreover, any number n in the sequence formed by adding the starting pair and each derived pair is given by the equation:

$$11^{n-1} \times (a + b)$$

Thus the first term is $11^0 \times 136 = 136$; the second term is $11^1 \times 136 = 1496$ and the third is $11^2 \times 136 = 16{,}456$, exactly as we found during the cross-adding.

Mirror equivalence

There are integers whose squares when reversed—that is, read from right to left instead of from left to right—become the squares of the reverse of the original number. For example, 12^2 is 144 and the reverse of 144 is 441, which happens to be the square of 21, the reverse of 12. This feature is remarkable enough to warrant asking how common is it. And we might also ponder how it is distributed among all the integers: does it clump or is it evenly spread?

To help in exploring this group of integers, let's create some relevant variables:

s = the starting integer

s^2 = the square of the starting integer

t = the mirror image of s^2 (that is, the digits placed in the reverse order), such that s^2 and t have the same number of digits

u = the square root of t.

Thus for some s, there is a corresponding u such that u is the mirror of s. Let's call this *mirror equivalence*.

We can also represent the process of generating mirror equivalence in this way:

$$s \rightarrow s^2 \rightarrow \mathit{Я}(s^2) \rightarrow \sqrt{\mathit{Я}(s^2)} \rightarrow \mathit{Я}(s)$$

where s is the starting term and $\mathit{Я}$ is the operation of reversing the digits of the accompanying argument.

There are three trivial instances of mirror equivalence: 1, 2 and 3.[17] But it becomes more interesting when the squares, and their mirror images, are multi-digit. After much doodling, I have found that there are two types of mirror equivalence:[18]

- Type 1: the square of the starting number is a palindrome—that is, reads the same from either end—and thus the square root of the mirror palindrome must be the mirror of the starting number. For example, if $s = 22$, s^2 is the palindrome 484, and the square root of the mirror of this palindrome is 22, the mirror of the starting term s. In the first 100,000 integers, there are 21 such numbers that produce a palindrome on squaring. I'll call integers of this type *palindromic mirrors*.

- Type 2: the square of the starting number is not a palindrome and yet the square root of the mirror square is a mirror of the starting term. For example, if $s = 13$, s^2 is 169, $t = 961$ and $u = 31$, the mirror of s. In the first 100,000 integers, there are just 132 such numbers. I'll call integers of this type *non-palindromic mirrors*.

---

17. For example, $3^2 = 9$, 9 reversed is 9, the square root of 9 is 3 and 3 is the mirror of 3 (in the sense of being the same as the starting number reversed).

18. If you want to explore this using Microsoft Excel, you can create a custom function to quickly reverse the digits of any number. See "Reversing digits" on page 178.

Clearly, there are not that many integers that exhibit mirror equivalence. It is quite rare: just 0.153% of the first 100,000 integers. I have listed them in appendix D, starting on page 183, segmented by type.

The palindromic mirrors are perhaps the least interesting, although it may be worth noting that they always seem to be palindromes themselves (for example, 101, 111, 121 and so on).[19] What seems more remarkable is the case where the square of the starting number is not a palindrome and yet the square root of the mirror square is a mirror of the starting term. We called these non-palindromic mirrors and noted there are just 132 of them among the first 100,000 integers. They are a rare breed indeed. They form the sequence that begins {12, 13, 21, 31, 102, 103, 112 ...}. But we can add another curious feature to their rarity. Of all the starting integers subjected to our *square–reverse–square-root* process (hereafter SRSR) that produce a resulting integer of the same number of digits, the non-palindromic mirrors are the most numerous.[20] For example, in the range 1 to 10,000, there are only 73 equi-digit integers produced by the SRSR process, 50 of which are non-palindromic mirrors. In the range 10,001 to 20,000, only 53 equi-digit integers are produced by SRSR, 47 of which are non-palindromic mirrors. These remarkable numbers show abundance within scarcity. They are a copse within a forest.

The SRSR process generates some other interesting cases. We have already mentioned the palindromic

---

19. You should not conclude from this that only a palindrome produces a palindrome on squaring. We'll see some counterexamples a little later.

20. We can ignore those starting integers that produce a integer with a different digit length since they cannot be a candidate for mirror equivalence. For example, $10^2 = 100 = 1$ when reversed = the square of 1. Clearly, 10 is not the mirror of 10.

mirrors, but there are other integers worthy of mention. For example, there are cases where u (the final number) is an integer, but it is not the mirror of s (the starting number). For example, if s is 26, u is also 26—and clearly 26 is not a mirror image of 26. However, 26^2 is a palindrome: 676. Let's call these type 3 integers *palindromic non-mirrors*. There are also cases were u, though not the mirror of s, is an integer of the same length as s and s^2 is not a palindrome. For example: if s is 33, s^2 is the non-palindrome 1089 and u is 99. Let's call these type 4 integers *non-palindromic non-mirrors*.

As well as the mirror-equivalent integers, I have also listed the non-mirroring integers (those of type 3 and 4) in appendix D. They too are rare. In the first 100,000 integers, there are only 9 type-3 integers and 6 type 4. Thus there are just 168 integers in the first 100,000 that, through the SRSR process, generate an integer with the same number of digits. A thousand generate an integer without the same number of digits (such as 10, which generates 1, and 120, which generates 21). Thus 99.832% of all the integers as far as 100,000 do not generate a whole number when subjected to SRSR. This again highlights the rarity of mirror equivalence.

You might think, given the apparent rarity of mirror equivalence, that the number of integers that generate non-mirroring would be significantly greater. But as we've just seen, that is not so. In the first 100,000 integers, there are just 15 non-mirroring (or type 3 and 4) integers: 26, 264, 307, 836, 2285, 2636, 22,865, 24,846 and 30,693 being palindromic non-mirrors, and 33, 99, 3168, 6501, 20,508 and 21,468 being non-palindromic non-mirrors.That should strike you as odd. The situation where there is *no* restriction on the digital ordering of the result is more than 10 times *less* likely to occur than the situation where a strict restriction is placed on the order of digits!

The relative scarcity of mirror equivalence (and of types 3 and 4 integers) might seem to have a simple explanation. To return to the symbols we used on page 96, for u to be an integer, t must be a perfect square (since u is the square root of t). But the number of perfect squares within any range of integers (r) is always less than r. For example, in the range 1 to 5000, there are only 70 perfect squares. So there cannot be as many mirror equivalences as there are starting integers. Further, the number of perfect squares thins out the further we get from zero. Between 5000 and 10,000, there are 29 perfect squares, less than half the number between 1 and 5,000. There are even some equivalent ranges where there are *no* perfect squares. One example among many: between 7,885,000 and 7,890,000. Thus the density of mirror equivalences should also thin out the further we get from zero, since mirror equivalence can only grow out of perfect squares. And that is what seems to be the case (as suggested by Table 4 on page 101).

Now the doodler, desperately trying to steal the odd moment of arithmetic play from the quotidian humdrum of work–sleep–eat, might well be forgiven for taking Table 4 as strong inductive evidence that mirror equivalence is finite. (Indeed, our listing in appendix D of those integers in the range of 1 to 100,00 that exhibit mirror equivalence ends at 31,111.) But that would be a mistake. Despite the apparent thinning out of perfect squares as the integers grow, mirror equivalence pops up again and again. For instance, there are 58 mirror-equivalent integers in the range 100,001 to 110,000, 48 between 110,001 and 120,000, 19 between 120,001 and 130,000, and 4 between 130,001 and 140,000. Then there is no more mirror equivalence until 200,001 to 210,000, where we find 34 instances!

Table 4: Diminishing mirror equivalence per starting number range

Starting integer range	No. of mirror equivalences
1–10,000	63
10,001–20,000	53
20,001–30,000	29
30,001–40,000	8
40,001–50,000	0
50,001–60,000	0
60,001–70,000	0
70,001–80,000	0
80,001–90,000	0
90,001–100,000	0

Curiouser and curiouser. Clearly there is more to mirror equivalence than the distribution of perfect squares.[21]

Returning to the rarity of those numbers we have squeezed out of our SRSR mincer, consider finally those integers we labelled type 4, the non-palindromic non-mirrors. We noted that there are just 6 in the first 100,000 integers. In the next 100,000 there is just one more, and in the 100,000 integers after that, there is just one more again,

---

21. There is a parallel here with prime numbers. There are some large gaps between consecutive prime numbers, sometimes as large 1475, and perhaps larger. But since there is an infinite number of prime numbers, we know that, however large a gap, another prime will eventually pop up. (We discuss the infinity of prime numbers on page 106.) Might not this be the same with mirror-equivalent integers?

making the sequence of non-palindromic non-mirrors between 1 and 300,000:

{33, 99, 3168, 6501, 20,508, 21,468, 110,922, 219,111 ...}.[22]

Having reached the limit of my computing power, I now pose some challenges.

Challenge 1

Is the sequence of non-palindromic non-mirrors finite?

Challenge 2

Is there an integer s such that:

s^3 = the cube of s

t = the mirror image of s^3 (that is, the digits placed in the reverse order), such that s^3 and t have the same number of digits

u = the cube root of t and

the mirror image of u is s.

There are two trivial answers: 1 and 2. But are there any others?

Prime numbers: The quarks of our number system

Prime numbers have fascinated mathematicians, professional and amateur, for thousands of years. They seem somehow to be fundamental to our natural number system, the system of counting numbers: the integers from 1 onwards. Every number is itself a prime number or can, in various ways, be derived from prime numbers. That's

---

22. Note that the digital root of each number in this sequence is 3, 6 or 9.

what drives mathematicians to call prime numbers the *fundamental building blocks* of number systems. They are to numbers what quarks are to protons, neutrons and electrons—the elementary particles that together compose the atoms from which all matter derives.

But despite the central importance of prime numbers, they are an unruly lot. Mathematicians have struggled for centuries to derive some formula or equation that can generate the prime numbers. But they have had no success. There seems to be no rhyme nor reason, no pattern obvious or otherwise, in the progressive list of prime numbers. You cannot plug, say, 215 into some formula and hope to get out of it the 215th prime number, 900 and get the 900th prime number, 2134 and get the two thousand, one hundred and thirty-fourth prime number, and so on.

So what are they?

A prime number is a positive integer greater than 1 that has no positive factors except itself and 1. Put another way, a number is a prime number if no other numbers divide into it evenly—that is, without there being a remainder—other than 1 and itself.

It follows that 7 is a prime number because no other number divides into it without there being a remainder. For example, $7 \div 2$ gives 3 plus a remainder of 1; $7 \div 3$ gives 2 plus a remainder of 1; $7 \div 4$ gives 1 plus a remainder of 3, $7 \div 5$ gives 1 plus a remainder of 2; and $7 \div 6$ gives 1 plus a remainder of 1. Since no number divides into 7 without there being a remainder—other than 1 and 7— 7 therefore is a prime number.

On the other hand, 8 is not a prime number, since at least one other number divides evenly into it. For example, 2 divides evenly into 8 (giving 4 with no remainder). Numbers that are not prime are known as *composite numbers* (or, more strictly, *positive integers* divisible by a number other than themselves or 1 are known as composite numbers).

The first 10 prime numbers are 2, 3, 5, 7, 11, 13, 17, 19, 23 and 29. (All the prime numbers less than 10,000 are given in appendix B starting on page 171. There are 1210.) The first composite numbers in the same range are therefore 4, 6, 8, 9, 10, 12, 14, 15, 16, 18, 20, 21, 22, 24, 25, 26, 27 and 28.

Note that even though 1 is divisible only by 1 and itself, it is not—or, more accurately, no longer—considered a prime number. This is largely the result of definitional fiat. 2 was once not considered a prime, but nowadays it is. It is the only even prime. (Every even number greater than 2 is obviously divisible by 2; hence no prime number greater than 2 can be an even number.)

In the beginning was Euclid

Euclid, the famous Greek mathematician who flourished around 300 BCE, established a number of important facts about prime numbers. For example:

- every composite number must have a factor that is a prime number[23]
- there is an infinite number of prime numbers.[24]

Let's consider how Euclid derived these facts before considering some other curious features of prime numbers.

Every non-prime number has a prime factor

Euclid's proof that every composite number—that is, every positive integer that isn't a prime number—has factors that are prime numbers is quite simple. If a

---

23. Euclid, *Elements*, Book VII, Proposition 31.
24. *ibid.*, Book IX, Proposition 20.

number, *A*, is composite, then it must have factors other than *A*, that is, there must be at least one number other than *A* that divides evenly into it. (Remember that only prime numbers don't have such factors, and composite numbers are defined as those that are not prime numbers.)

Suppose that *B* is a factor of *A*. If *B* is a prime number, then QED. But if *B* is composite, then it must, by definition, have at least one factor other than itself and 1.

Suppose that *C* is a factor of *B*. Then if *C* is a prime number, then QED. But if *C* is composite, then it too has at least one factor other than itself and 1.

By now the prospect of an infinite number of factors of the finite number *A* has reared its ugly head. Since this is impossible, somewhere in the chain of factors there must be a factor that is a prime number.

Let us suppose, for the sake of argument, that we find that *E* is a prime number. Let us suppose, too, that *E* is a factor of *D* and *D* a factor of *C* (which we have established is a factor of *B* and *B* is a factor of *A*). It follows that *E* must be a factor of *C*, *B* and *A*. Hence the composite number *A* has a prime number as a factor.

If the proof seems a little vague, an example might help illuminate it. Take any composite number, say 768, and think of a factor of it, say 48. (Remember that a factor of a number is simply another number that divides evenly into it.) Keep doing likewise. Regardless of the factors you choose, you will end up with at least one prime number in your chain of factors. In the figure shown at the top of page 106, the chosen chain of factors has ended with 3, a prime number. (Indeed, if you choose any other path, you will end up with 2, also a prime number. For instance, 16 is a factor of 768. Now choose 8, 4 and, inevitably, 2.)

768 (composite)

48

24

6

3 (prime)

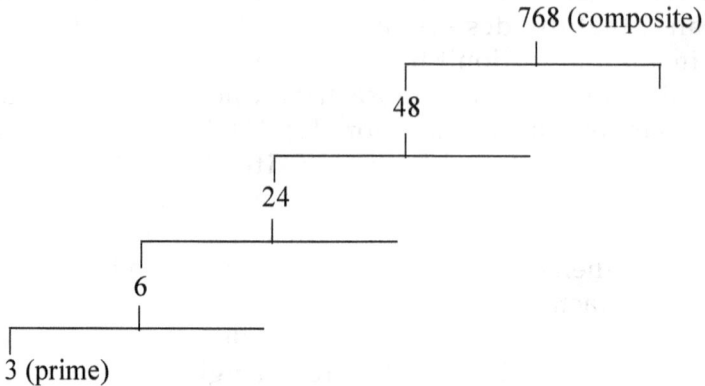

Figure 1: Repeatedly factorising a composite always yields a prime

There is an infinite number of primes

Having established that every number is either a prime or has a factor that is a prime, Euclid goes on to prove that there must be an infinite number of primes. The argument goes like this:

Suppose that there is only a finite number of primes: A, B, C, D ... If you multiply all these prime numbers together (to get a number N) and add one to the result (that is $N + 1$), the question arises as to whether $N + 1$ is a prime number.

Now $N + 1$ is obviously greater than any one of the prime numbers (A, B, C, D ...) that, when multiplied together, generated N. But if $N + 1$ is not one of the numbers A, B, C, D ... and the numbers A, B, C, D ... constitute *all* the prime numbers, then $N + 1$ cannot be a prime number.

If $N + 1$ is not a prime number, it must be a composite number. Now every composite number has a factor that is a prime number. (We established this in the last section.)

Let's assume that G is a factor of $N + 1$ and G is a prime number. Euclid now argues that G cannot be the same as any of the primes in the original finite set of primes: A, B, C, D ... For if G is one of the finite prime numbers A, B, C, D... then it would follow that G divides evenly into both N and into $N + 1$. (It divides evenly into N because N is just the product of all the prime numbers of which G is one.) But if G divides evenly into both N and into $N + 1$, G must divide evenly into 1, which is not possible if G is a prime number. (Remember that 1 is not a prime number.)

It follows, then, that G is a prime number that is not one in the set of supposedly finite prime numbers. Therefore, there are more prime numbers than any limited or finite set of prime numbers. There are, in other words, an infinite number of them.

If the proof seems a little vague, an example might help illuminate it. Suppose we are told that there are only 5 prime numbers: 2, 3, 5, 7 and 9. We multiply all these numbers together to get 1890, and then add 1 to get 1891. If 1891 is a prime number (which it isn't), we have a prime number other than 2, 3, 5, 7 and 9. But if 1891 is a composite number, it will have a factor that is a prime number. If 2, 3, 5, 7 and 9 are the only prime numbers, then at least one of these should divide evenly into 1890 (which obviously they all do) and into 1891. But if there is a prime number in our allowable list of prime numbers that divides evenly into 1890 *and* 1891, then it must be able to divide evenly into 1, which none can. (Another way of putting this is that since the smallest prime number is 2, if a prime number can divide evenly into a number N, then the next possible number that any prime number could divide evenly into is $N + 2$, and only then if the prime number is 2.) So there has to be a prime factor of 1891 that is neither 2, 3, 5, 7 or 9. And a little mental arithmetic will show that 31, a prime number, divides evenly into 1891.

The elementary particles of our number system

Let's now go a little further than introductory Euclid. Prime numbers are sometimes called the

Every positive integer is either a prime number or the product of one set, and one set only, of prime numbers.

atoms, the building blocks, or the elementary particles of the number system. This is because every positive integer (except 1) is either a prime number or a product of prime numbers. This is part of what is known as the *fundamental theorem of arithmetic*. But this theorem goes further and states that a composite number can be expressed as the product of primes *in only one way*. In other words, for any composite number, there is a *unique* collection of prime numbers that, when multiplied together, gives that composite number.[25]

For example, the composite number 60 is the product of four prime numbers: $2 \times 2 \times 3 \times 5$. According to the fundamental theorem of arithmetic, no other combination of prime numbers will yield 60 when multiplied together. Consider another example: $102 = 2 \times 3 \times 17$. Now 2, 3 and 17 are prime numbers, and, as a spot of mental arithmetic will prove, no other combination of prime numbers will, when multiplied together, give 102. Of course, 102 has other factors: $51 \times 2 = 102$, but this is not a disproof of the fundamental theorem of arithmetic, since 51 is not a prime

---

25. For this reason, the fundamental theorem of arithmetic is sometimes called the *unique factorisation theorem*. Incidentally, the fundamental theorem of arithmetic follows from another of Euclid's theorems, namely that if a prime number divides into a composite number, then it also divides into at least one of its factors. Note that this so-called fundamental theorem of arithmetic would not be true if 1 was considered a prime number (which it would be if the common definition of a prime number was true, namely, a number divisible only by itself and 1). For then there would be composite numbers that could be expressed as the product of primes in more than one way. For example, $15 = 5 \times 3$ and $5 \times 3 \times 1$.

number. It is, in fact, the product of 3 and 17. Consider some more examples:

61 = a prime number

62 = 2 × 31

63 = 3 × 3 × 7

64 = 2 × 2 × 2 × 2 × 2 × 2

65 = 5 × 13

66 = 2 × 3 × 11

and so on. In every case, the number is either a prime number (as in the case of 61) or the product of a unique set of prime numbers.

§

We have seen that prime numbers are, in certain ways, the building blocks of the whole number system, for every whole number is either indivisible (and thus a prime number) or is composed of indivisibles multiplied together. There are other ways in which prime numbers might be considered the building blocks of whole numbers. In fact, it may even be the case—as I'll conjecture later—that prime numbers themselves (or at least those greater than 7) can be broken down into particular *combinations* of other prime numbers.

Goldbach's conjecture and some variations

In a 1742 letter to the famous Swiss mathematician Leonard Euler, the Prussian mathematician Christian Goldbach suggested that every even integer greater than 5 can be expressed as the sum of three prime numbers.

Euler provided a slightly modified version of Golbach's conjecture: that all positive even integers greater than 2 can be expressed as the sum of two primes. It is Euler's version that is now referred to as *Goldbach's conjecture*.

Although Goldbach's conjecture has been verified for numbers up to 4×10^{18}, it has never been formally proved.[26]

A variation on Goldbach's conjecture, known as the *Odd Goldbach conjecture*, states that every odd integer greater than 5 can be expressed as the sum of three primes. (For example, $23 = 3 + 7 + 13$, $25 = 5 + 7 + 13$, $27 = 7 + 7 + 13$, etc.) Like the original Goldbach conjecture, the Odd Goldbach conjecture has never been proved, although it has been shown to be true for a vast number of odd integers.[27]

And now for some further prime conjectures

The fundamental theorem of arithmetic states that all whole numbers can be expressed, and expressed uniquely, as the *product* of prime numbers. Goldbach conjectured that all numbers can be expressed as the *sum* of two (or three) prime numbers. Inspired by these conjectures, I'd like to propose a few of my own.

---

26. The 4×10^{18} figure was as of 2012. See http://sweet.ua.pt/tos/goldbach.html. Viewed 11 February 2017.

27. See http://www.utm.edu/research/primes/notes/conjectures/. Viewed 11 February 2017.

Conjecture 1

If we accept that 1 is not a prime number, then our first conjecture is that all whole numbers greater than 5 can be expressed as:

$$p + (q \times r)$$

where p, q and r are prime numbers. (Indeed, if 1 were a prime number, then this expression would, I suggest, hold for *all* whole numbers greater than 1.)

Table 5 gives values for p, q and r for a sample of numbers. I have found no exceptions after analysing many thousands of whole numbers.

Table 5: Conjecture 1

Whole number	*p*	*q*	*r*	*p* + (*q* × *r*)
11	2	3	3	11
24	3	3	7	24
63	5	2	29	63
100	13	3	29	100
101	7	2	47	101
102	37	5	13	102
103	17	2	43	103
515	13	2	251	515
911	13	2	449	911
1009	2	19	53	1009
1051	929	2	61	1051
2000	17	3	661	2000
3917	139	2	1889	3917
7500	17	7	1069	7500

The number in the final column of Table 5 is obtained by adding a prime number (p) to the product of two other prime numbers (q and r). So, for example, $103 = 17 + (2 \times 43)$. If you refer to appendix B on page 171 you will see that all the values in columns p, q and r are indeed prime numbers.

The conjecture appears to be a general one and not limited to the relatively few prime numbers listed in appendix B. For example, 1,000,000 can be expressed as 23 + (11 × 90,907) and 90,907 is a prime number.

Note that most whole numbers can be generated from more than *one* set of values for p, q and r. For example, Table 6 shows the eight ways that 102 can be decomposed into three prime numbers—p, q and r—according to the formula $1009 = p + (q \times r)$.

Table 6: Example of multiple decompositions (conjecture 1)[a]

p	q	r	p + (q × r)
93	3	3	102
67	7	5	102
53	7	7	102
47	11	5	102
37	13	5	102
17	5	17	102
11	7	13	102
7	19	5	102

a. If we reverse the q and r values, there are 14 ways, not 8.

In other words, I am not claiming that each whole number can be derived from a *unique* set of constituent prime numbers according to the formula $p + (q \times r)$. The claim is merely that there will always be *at least one* set of constituent prime numbers. Indeed, it seems reasonable to conjecture that as the numbers being decomposed into constituent primes get larger and larger, the greater is the number of possible unique solutions. For example, while there are eight ways of decomposing 102, there are 184 ways of decomposing 1009 into $p + (q \times r)$ where p, q and r are prime numbers.

Finally, note that for any odd number greater than 5, either:

- $p = 2$ or
- $q = 2$, $r = 2$ or both q and $r = 2$.

We can see this from the following exhaustive table:

Case	If $p + (q \times r)$ is ...	... and p is ...	$(q \times r)$ must be ...
1	odd	even (and thus 2)	odd
2		odd	even
3	even	even (and thus 2)	even
4		odd	odd

For the odd numbers (cases 1 and 2), if p is not an even prime number (that is, not 2), $(q \times r)$ must be even. This is so because only when an even number is added to an odd number do we get an odd number. Thus p, q or both must be 2, for there is simply no other way of making $(p \times r)$ even. (A modicum of mental arithmetic will show that none of 1, 3 or 5 can be expressed as $p + (q \times r)$ where p, q and r are prime numbers).

Note that case 2 is Lemoine's conjecture, namely, that every odd integer is the sum of an odd prime and an even semi-prime (where a semi-prime is simply the product of two primes). In effect, Lemoine's conjecture becomes this: every odd number greater than 5 can be expressed as an odd prime number plus *twice* another prime number, a conjecture that was also claimed by Cambridge University mathematicians Godfrey Hardy and John Littlewood in 1922[28], not to mention Hyman Levy in 1963.[29]

Ideas for further exploration

Are there numbers with only one decomposition into $p + (q \times r)$, where p, q and r are prime numbers?

Conjecture 2

Our second conjecture seems rather more extraordinary. But, once again, many hours of computer analysis shows that it is certainly the case for many thousands of prime numbers, and this should give some encouragement to the belief that the conjecture holds for *all* prime numbers. The conjecture is this:

All prime numbers greater than 7 can be expressed as:

$$p + (q \times r)$$

where p, q and r are prime numbers and

$$(p \times q) + r$$

is also a prime number.

Note that all we have done in part two of the conjecture is switched the arithmetic operators: addition in part one becomes multiplication in part two, and multiplication in part one becomes addition in part two. Note too that the first part of the conjecture follows from the first conjecture, discussed on page 111 (that is, if the

---

28. Laurent Hodges, "A Lesser-Known Goldbach Conjecture", *Mathematics Magazine*, vol. 66, no. 1, 1993, pp. 45–47.
29. Hyman Levy, "On Goldbach's Conjecture", *Mathematical Gazette*, iss. 47, 1963, p. 274.

first conjecture is true, then the first part of the second conjecture is also true).

Table 7 shows the results of conjecture 2 for a sample of prime numbers.

Table 7: Conjecture 2

Prime number	p	q	r	p + (q × r)	(p × q) + r
11	5	2	3	11	13
13	7	2	3	13	17
17	3	2	7	17	13
113	7	53	2	113	373
263	97	2	83	263	277
499	41	2	229	499	311
1319	2	439	3	1319	881
2711	997	2	857	2711	2851
3917	1279	2	1319	3917	3877
4271	1069	2	1601	4271	3739

In each instance, a particular dissection of a prime number into three constituent primes — p, q and r — according to the formula $p + (q \times r)$ is such that $(p \times q) + r$ is also a prime. Appendix B will show that *all* the numbers in the Table 7 are indeed prime numbers.

Again, the conjecture is a general one and not limited to the prime numbers listed in appendix B. For example, 90,907 can be expressed as $29 + (2 \times 45,439)$. Each constituent number is a prime. By swapping the arithmetic operators we get $(29 \times 2) + 45,439$ which equals 45,497, and 45,497 happens also to be a prime number.

There are many prime numbers for which there are a number of sets of constituent primes that satisfy the conjecture. Consider 53, for example. As Table 8 shows, there are four ways of decomposing 53 into $p + (q \times r)$ is such that $(p \times q) + r)$ is also a prime.

Table 8: Example of multiple decompositions (conjecture 2)

Prime number	p	q	r	$p + (q \times r)$	$(p \times q) + r$
53	2	3	17	53	23
53	2	17	3	53	37
53	7	2	23	53	37
53	7	23	2	53	163

But the conjecture is not that, for each prime, there is a *unique* set of primes p, q and r such that $p + (q \times r)$ equals that prime number while $(p \times q) + r$ is also a prime. The conjecture is simply that, for each prime number, there will always be *at least one* set of constituent primes for which those two conditions are true.

And just to show that this conjecture does not hold for all composite numbers (that is, non-primes) as well, consider all the ways of representing, say, 36 as $p + (q \times r)$ and $(p \times q) + r$ where p, q and r are prime numbers (shown in Table 9 on page 117).In this exhaustive list, there is no instance where $(p \times q) + r$ yields a prime number. This is why conjecture 2 applies solely to prime numbers (of which 36 is not).

However, conjecture 2 does not apply *exclusively* to prime numbers. There are some composite numbers that can be dissected into $p + (q \times r)$ such that $(p \times q) + r$ is also a prime number. For example, 16 is not a prime number, and yet 16 = 2 + (2 × 7) and (2 × 2) + 7 = 11, and 11 is a prime number. Likewise, 25 = 3 + (2 × 11) and (3 × 2) + 11 = 17, and 17 is a prime number.

Table 9: Example of a composite failing conjecture 2

Number	*p*	*q*	*r*	*p* + (*q* × *r*)	(*p* × *q*) + *r*
36	2	2	17	36	21
36	2	17	2	36	36
36	3	3	11	36	20
36	3	11	3	36	36
36	11	5	5	36	60

Conjecture 3

Conjecture 3 extends conjecture 2 by considering the one other way in which the arithmetic operators in $p + (q \times r)$ can be switched. The conjecture is that:

the *vast majority* of prime numbers can be expressed as:

$$p + (q \times r)$$

where p, q and r are prime numbers,

$$(p \times q) + r$$

is also a prime number and

$$(p \times r) + q$$

is a further prime number.

Table 10 on page 118 shows the results of conjecture 3 for a sample of prime numbers.

Table 10: Conjecture 3

Prime number	p	q	r	$p + (q \times r)$	$(p \times q) + r$	$(p \times r) + q$
11	5	2	3	11	13	17
13	3	2	5	13	11	17
37	2	5	7	37	17	19
89	43	23	2	89	991	109
401	7	2	197	401	211	1381
509	503	2	3	509	1009	1511
733	347	2	193	733	887	66,973
953	787	2	83	953	1657	65,323

Note that every number in this table is a prime number. (Refer to appendix B for primes less than 10,000, if necessary.[30])

There are just 14 exceptions among the 168 prime numbers up to 1,000: 59, 79, 181, 227, 251, 283, 431, 547, 571, 599, 647, 701, 829 and 983. This is 8.333% of the prime numbers in this range. There are seven more exceptions in the 135 prime numbers between 1,000 and 2,000: 1091, 1153, 1171, 1399, 1427, 1709 and 1847. This is 5.185% of the prime numbers in this range. In the 127 primes between 2000 and 3000, there are just two exceptions: 2309 and

---

30. There are numerous online prime-number checkers. One example is at http://www.math.com/students/calculators/source/prime-number.htm. Many handheld calculators have a prime-number checking function. For example, the *HP Prime* graphing calculator offers the isprime() function. I have provided the code for a Microsoft Excel function that will check if a number is prime (see page 179).

2617 (or 1.575% of the primes). It is tempting to conjecture further that the number of exceptions continues to reduce as the prime numbers become larger and larger. Indeed, it is likely that the larger the prime number, the greater the number of possible decompositions into constituent prime numbers according to our combinatorial arithmetic patterns. And with more possible decompositions likely to meet our conjecture, the less likely it becomes that an exception will be found. Which brings me to a further conjecture:

> Beyond a certain prime number, X, *every* prime number can be expressed as:
>
> $$p + (q \times r)$$
>
> where p, q and r are prime numbers,
>
> $$(p \times q) + r$$
>
> is also a prime number and
>
> $$(p \times r) + q$$
>
> is a further prime number.

What X is I'll have to leave to someone with a far better mathematical brain than mine.

Conjecture 4

The Lemoine conjecture — unwittingly re-proposed by Hardy and Littlewood 27 years later, and Levy 68 years later — is that every odd number greater than 5 is the sum of an odd prime and the double of a prime. Since every prime greater than 2 is an odd number, a corollary of the Lemoine conjecture is that every prime greater than 5 is the sum of an odd prime and the double of a prime. Large-scale computer iteration lends weight to the conjecture.[31]

What now of a seemingly related conjecture, namely:

Every prime number greater than 2 can be expressed as:

$$p - 2q$$

where p and q are prime numbers.

Table 11 gives some examples from a wearying iterative slog at testing for exceptions:.

Table 11: Conjecture 4

Prime number	p	q	$p - 2q$
3	7	2	3
23	29	3	23
97	103	3	97
229	233	2	229
647	673	13	647
887	1009	61	887
2683	4337	827	2683
3181	3187	3	3181
8941	8963	11	8941

A decomposition according to conjecture 4 is not necessarily unique. Take 229, for example. From the table above we see that it can be expressed as 233 − (2 × 2). But it can also be expressed as 239 − (2 × 5), 743 − (2 × 257), 1907 − (2 × 839), 4283 − (2 × 2027), and 9587 − (2 × 4679).

---

31. As at 1999, Lemoine's conjecture had been verified up to 10^9. See http://mathworld.wolfram.com/LevysConjecture.html. Viewed 6 November 2016.

Take another prime number, say, 887. It can be expressed as 1009 – (2 × 61). But it can also be expressed as 3061 – (2 × 1087), 3733 – (2 × 1423), 4021 – (2 × 1567) and 7129 – (2 × 3121). And one more, say, 2683. It can be expressed as 4337 – (2 × 827) and also as 9929 – (2 × 3623).

These multiple decompositions tempt one to suggest another conjecture: for every prime number greater than 2, there is an *infinite number* of decompositions of the form $p - 2q$ where q and q are also prime numbers. I'll leave the proof (or disproof) of that to sharper brains.

Although conjecture 4 appears to mirror the Lemoine conjecture, the potential infinity of decomposition possibilities offered by conjecture 4 cannot also apply in the case of the Lemoine conjecture. In the equation $N = p + 2q$, it should be clear that p and q must always be less than N. (Indeed p must be $\leq (N - 3)$ and q must be $\leq ((N - 3)/2)$.) Thus the theoretical limit on the number of possible decompositions is $((N - 3)/2)$.[32] Since this limit must always be less than N, there cannot be an infinite number of such decompositions. But in our equation ($N = p - 2q$), p must always be greater than N, leaving open the possibility of an infinite number of solutions.

Conjecture 5

It is easy to show that no *even* composite number (say E) can be expressed as $p - 2q$ where p and q are prime numbers. For $2q$ will always be even, and if E is even, p must also be even. Now the only even prime number is 2. But if $p = 2$, E will be a negative number (since $2q$ is obviously greater than 2 if q is a prime number). But composite numbers are, by definition, *positive* integers.

---

32. Of course, the actual limit will be less, as not all the theoretical decompositions will involve only prime numbers.

What about odd composite numbers? Our conjecture here is that all odd composite numbers can be expressed as $p - 2q$ where p and q are prime numbers. Table 12 gives some examples.

Table 12: Conjecture 5

Odd composite number	p	q	$p - 2q$
9	13	2	9
15	29	7	15
93	239	73	93
497	1039	271	497
749	9931	4591	749
1019	1861	421	1019
7319	8053	367	7319

Might there be an infinite number of such decompositions? Note that 9 can also be expressed as $19 - (2 \times 5)$, $23 - (2 \times 7)$, $31 - (2 \times 11)$, $43 - (2 \times 17)$ and $47 - (2 \times 19)$. And 15 is also equal to $37 - (2 \times 11)$, $41 - (2 \times 13)$, $53 - (2 \times 19)$, $877 - (2 \times 431)$ and $2621 - (2 \times 1303)$. And one more for good measure: 93 can also be expressed as $971 - (2 \times 439)$, $2131 - (2 \times 1019)$, $4259 - (2 \times 2083)$ and $9931 - (2 \times 4919)$. So, paralleling our secondary suggestion in conjecture 4 above, a feasible conjecture is that there is an *infinite number* of decompositions of odd composite numbers into the form $p - 2q$ where q and q are prime numbers. Again, I'll leave the proof (or disproof) of this conjecture to those whose minds are endowed with far greater mathematical prowess.

Conjecture 6

Every prime number greater than 2 can be expressed as:

$$2p - q$$

where p and q are prime numbers.

Table 13 gives some examples.

Table 13: Conjecture 6

Prime number	p	q	2p – q
3	5	7	3
23	47	71	23
179	263	347	179
811	547	283	811
1901	2141	2381	1901
4001	2141	281	4001
7321	4261	1201	7321

After churning through thousands of prime numbers, I have not encountered a single counter-example.

Here too there appear to be good grounds for hypothesising that there is an infinite number of decompositions of every prime number. For example, 3 can also be expressed as $(2 \times 127) - 251$, $(2 \times 181) - 359$, $(2 \times 1051) - 2099$, and so on. And 7321 can be expressed as $(2 \times 7027) - 6733$, $(2 \times 8161) - 9001$ and $(2 \times 8821) - 10321$. While these are just sample decompositions, the minus sign in our template encourages a wild hypothesis, especially given the infinity of numbers.

Conjecture 7

As we did with conjectures 4 and 5, we could ask whether the apparent relation in conjecture 6, (namely, $2p - q$) holds also for composite numbers. Let's start with the even composites. This possibility can be quickly discarded with a little mental arithmetic. For if the composite number is even, both sides of the subtraction would have to be even, which limits q to 2. The two even composites up to 4 can be decomposed as $2p - q$, but trouble starts with the very next even composite: 6. $2p$ would have to be 8, making p a composite, not a prime, number.

What about odd composite numbers? This looks more promising, as q would now have to be odd, and the overwhelming majority of primes are odd numbers. Indeed thousands of computerised iterations prompt the following conjecture:

Every odd composite number can be expressed as:

$$2p - q$$

where p and q are prime numbers.

Table 14 on page 125 gives some examples.

Again, there are good grounds for proposing a related conjecture, namely, that there is an infinite number of decompositions for each odd composite number. For example, 9 can also be expressed as $(2 \times 311) - 613$, $(2 \times 563) - 1117$, $(2 \times 83) - 157$, and so on. 763 can also be expressed as $(2 \times 571) - 379$, $(2 \times 643) - 523$, $(2 \times 523) - 283$, and so on.

Table 14: Conjecture 7

Odd composite	p	q	2p − q
9	13	17	9
15	11	7	15
21	13	5	21
25	19	13	25
27	17	7	27
33	19	5	33
51	37	23	51
309	157	5	309
763	383	3	766

Conjecture 8

A number of the previous conjectures are based on an arithmetic schema of the form $p + (q \times r)$. We explored variations on the first form of this model, but all kept to the general form of adding one prime number to the product of two prime numbers. This initial schema appeared capable of generating all other numbers (and variations of it appeared to be able to generate the prime numbers).

Our seeming success with the $p + (q \times r)$ schema spurs us to consider a slight variation on it: $p − (q \times r)$. A spot of computer analysis does indeed reveal a similar set of observations to those that inspired conjecture 1. This prompts a new conjecture, namely:

All whole numbers can be expressed as:

$$p - (q \times r)$$

where p, q and r are prime numbers.

With conjecture 1, we found that the whole numbers between 1 and 5 escaped the $p + (q \times r)$ net. But all whole numbers from, and including, 1 seem to be capable of being expressed as $p - (q \times r)$. Table 15 gives values for p, q and r for a sample of such numbers.

Table 15: Conjecture 8

Whole number	p	q	r	$p - (q \times r)$
1	5	2	2	1
2	37	5	7	2
5	11	2	3	5
51	61	5	2	51
100	229	3	43	100
269	283	2	7	269
493	499	2	3	493
1765	1811	2	23	1765
3493	3499	2	3	3493
5000	5009	3	3	5000
7286	7321	5	7	7286

Again, a few sample numbers does not give significant weight to the conjecture, but I have not found a whole number that cannot be expressed as $p - (q \times r)$, where p, q and r are prime numbers, after examining many thousands of such numbers.

Note too that the decompositions are not always unique. Take 269 for example. Table 15 above shows that it can be decomposed into $283 - (2 \times 7)$. But it can also be decomposed into 307 − (2 × 19) and each number in the decomposition is a prime number. Perhaps, given the minus operator in our schema, we can confidently conjecture that there is an infinite number of ways that any whole number can be decomposed according to the schema.

Conjecture 9

Conjecture 8 grew out of an analysis of a variation on the arithmetic schema used in conjecture 1, namely $p + (q \times r)$. Conjecture 2 concerned prime numbers treated, firstly, according to that same schema and, secondly, according to a variation on that schema: $(p \times q) + r$. Now the parallel between conjectures 1 and 8 uncovered in the last conjecture prompts us to consider whether there might also be a parallel between conjecture 2 and another conjecture, one similar to conjecture 2 concerning the prime decomposition of prime numbers. And much iterative analysis of prime numbers suggests that this is the case, namely:

All prime numbers greater than 3 can be expressed as:

$$p - (q \times r)$$

where p, q and r are prime numbers, such that:

$$(p \times q) - r$$

is also a prime number.

Table 16 on page 128 shows the results of this conjecture for a sample of prime numbers. Again, the minus sign in the equation gives us grounds for thinking that there could well be an infinite number of ways any prime number could be decomposed in this way.

Table 16: Conjecture 9

Prime number	p	q	r	$p - (q \times r)$	$(p \times q) - r$
5	11	2	3	5	19
13	23	2	5	13	41
29	43	2	7	29	79
97	103	3	2	97	307
953	1627	2	337	953	2917
2087	2293	2	103	2087	4483

Conjecture 10

All whole numbers can be expressed as:

$$(q \times r) - p$$

where p, q and r are prime numbers.

Table 17 on page 129 gives values for p, q and r for a sample of numbers. I have found no exceptions after analysing many thousands of whole numbers.

What can be shown beyond doubt is that every *prime number* can be expressed as $(q \times r) - p$, where p, q and r are themselves prime numbers. The proof is simple, if not a little trivial: let N be a prime number. N is obviously equal to $2N - N$. So if $q = 2$ (which is a prime number) and r and p = N (thus making r and p necessarily prime numbers), $(q \times r) - p$ reduces to $2N - N = N$. For example, take 37, a prime number. Now $37 = (2 \times 37) - 37$. So whatever prime number you start with, you can dissect it into the product of itself and 2, minus itself. Thus every prime number can be expressed as $(q \times r) - p$ where p, q and r are prime numbers.[33] Where conjecture 10 goes beyond this particular proof is in claiming that every integer, whether prime or composite, can be expressed in the same way.

Table 17: Conjecture 10

Whole number	p	q	r	$(q \times r) - p$
1	3	2	2	1
2	7	3	3	2
5	5	2	5	5
69	5	2	37	69
100	11	3	37	100
490	499	23	43	490
997	997	2	997	997
2976	997	29	137	2976
9500	1601	17	653	9500

Conjecture 11

Our first conjecture (see page 111) is that all whole numbers greater than 5 can be expressed as:

$$p + (q \times r)$$

where p, q and r are prime numbers.

Let's modify this template a little and consider whether whole numbers can be expressed as:

$$p + (7 \times n)$$

where n is a positive integer and p is a prime number.

---

33. Of course, a more interesting conjecture is one where p, q and r are *unique* prime numbers.

It would seem that this conjecture has merit for all whole numbers greater than 29 (or greater than 28 if we allow n to be 0). If we allow 1 to be a prime number (as it once was), then our new conjecture seems to hold for all whole numbers greater than 14 (or greater than 7 if n can be 0). Here are some examples.

Table 18: Conjecture 12

Whole number (N)	n	p	$p + (7 \times n)$
30	4	2	30
101	4	73	101
479	10	409	479
812	115	7	812
1671	106	929	1671
14,625	32	14,401	14,625
254,123	892	247,879	254,123

You might expect that the number of ways a whole number (N) can be represented as $p + (7 \times n)$ would increase as N increases. That appears to be the case: 100 has just 4 representations, while 400 has 11. But there is one exception. When N is divisible by 7, there is only one way, namely:

$$N = 7 \cdot \left(\frac{N}{7} - 1 \right) + 7$$

For example, 252 divided by 7 is 36, and the only way to represent 252 according to our formula is:

$$252 = (7 \times (36 - 1)) + 7$$

Conjecture 12

Let's modify the template we used in conjecture 11 and consider whether whole numbers can be expressed as:

$$(7 \times n) - p$$

where n is a positive integer and p is a prime number.

It does seem that every whole number from 1 onwards can be represented in this way. Here are some examples.

Table 19: Conjecture 13

Whole number (N)	n	p	(7 × n) – p
30	5	5	30
101	16	11	101
319	46	3	319
479	1404	9349	479
812	117	7	812
1671	250	79	1671
14,625	2152	439	14,625
254,123	37,008	4933	254,123

As in conjecture 11, when N is divisible by 7, there is only one way to represent N according to our template. In this case, the template is:

$$N = 7 \cdot \left(\frac{N}{7} + 1\right) - 7$$

For example, 84 divided by 7 is 12, and the only way to represent 12 according to our formula is $84 = (7 \times (12 + 1)) - 7$.

§

We have not exhausted all possible arithmetic combinations of 2 and 3 prime numbers in our doodled decompositions. For example, what of $2p - 3q$? Are all prime numbers (or composite numbers) decomposable into that arithmetic schema? Prime numbers greater than 3 appear to be so. For example, $13 = (2 \times 11) - (3 \times 3)$ and $17 = (2 \times 13) - (3 \times 3)$. And our schemas $p + 7n$ and $7n - p$ seem to be just as fruitful when the 7 is replaced with certain other prime numbers (perhaps all?). However, I suspect that it is time to leave these explorations, lest their similarity induces a soporific effect on readers. We'll stick with primes for the time being, but move the spotlight a little. Still, it should be clear that there are many avenues open to the mathematical doodler to explore the decomposition of numbers—prime or otherwise.

Curiously repeating digits

Consider two integers, m and n, such that n is not a factor of m. Divide m by n to get p and consider just the fractional part of p. (For example, if $m = 100$ and $n = 12$, p is 8.3333... and the fractional part of p is .3333...). Now multiply p by 1, 2, 3, 4 and so on in consecutive order. You will find that the fractional part of each result eventually repeats, setting off an infinitely repeating cycle. (In the example of 100/12, the fractional parts as p is continually multiplied are .3333..., .6666..., .0000..., .3333..., .6666... etc., with the fractional part repeating after every three iterations.)

And now for an interesting conjecture:

> Where n is a prime number greater than 2, the number of iterations needed before the fractional part repeats is n.

For instance, divide 156 by the prime number 29. The fractional part is 0.379310344... and it will start repeating after the 29th iteration.

My excitement at having discovered this relationship was soon eclipsed by the realisation that this is not a feature restricted to prime numbers. Take, for example, 39112/69. Now 69 is not a prime number but the fractional part only repeats after the 69th iteration. Nor is this a feature of all numbers. In 126/44, the fractional part repeats after just the 22nd iteration. Indeed, looking closely at a large set of numbers prompts another conjecture:

> For any two numbers m and n such that n is not a factor of m, the fractional part of m/n will first repeat after either n iterations or f iterations, where f is a factor of n.

Note in the example above that 22 is a factor of 44. One more example: 2136/27 has a fractional part that repeats after every 9 iterations, and 9 is a factor of 27.

A little algebra will show that these are not terribly curious results.[34] Consider, firstly, that for every n—whether prime or composite—the fractional part of m/n will repeat after n iterations. Since our iterative algorithm is simply to multiply m/n (what we termed earlier as p) by consecutive integers, then the very next result after n iterations is:

---

34. Algebra will also help remove the uncertainty that inductive mathematical research often cannot avoid. For when we are dealing with decimal expansions, all mathematical tools available to us provide only limited numeric precision. For example, if you are using Microsoft Excel to compare decimal expansions—as I am—you need to be aware that Excel provides numeric precision only to 15 significant figures. Thus two otherwise identical decimal expansions might not appear to be identical in Excel, especially around the 14th, 15th and 16th figure.

$$\frac{m}{n} \times (n + 1)$$

which reduces to:

$$m + \frac{m}{n}$$

the fractional part of which is clearly the same as the fractional part of the very first iteration (being $m/n \times 1$). Similar reasoning will show that for $n + 2$, the fractional part is $2 \times (m/n)$, for $n + 3$, it is $3 \times (m/n)$ and so on, thus repeating the cycle. So the fractional parts always cycle after n iterations (and $2n$, $3n$ and so on).

But for some numbers, the repetition occurs before the nth iteration. We conjectured that this occurs after f iterations if f is a factor on n. That was a good first try. But it doesn't always work for *all* factors of n. Consider 112/60. The smallest factor of 60 (other than 1) is 2. But the repetition of fractional parts does not start until after the 15th iteration, not the second. Perhaps a little algebra can help us sharpen up our conjecture. Let n be equal to $f \times g$. In other words, f and g are two factors of n. Let's assume that after the fth iteration, the fractional part of $m/(f \times g)$ begins to repeat. The $f + 1$ iteration can be expressed as:

$$\frac{m}{f \times g} \times (f + 1)$$

which reduces to:

$$\frac{m}{g} + \frac{m}{f \times g}$$

Now $m/(f \times g)$ began our series of iterations. Clearly, if iteration $f + 1$ is to yield the same fractional part as the first iteration, the left side of the equation above must be an integer. If so, g must be a factor of m (that is, divide evenly into *m*). Thus we can fine-tune our conjecture as:

> For any two numbers *m* and *n* such that *n* is not a factor of *m*, the fractional part of *m/n* will first repeat after either *n* iterations, or after *f* iterations where *f* is a factor of *n* and *n/f* is a factor of *m*.

But even this refinement is not enough. Consider, for example, 264/60. One factor of 60 is 10, making the other factor (*g*) 6. Now 6 divides evenly into 264 — meeting our requirement that repetitions of fractional parts will begin before the 60th. But the *first* repetition occurs not after the 10th iteration, but after the 5th. So a further refinement is needed, namely:

> For any two numbers *m* and *n* such that *n* is not a factor of *m*, the fractional part of *m/n* will first repeat after either *n* iterations, or after *f* iterations where *f* is the *smallest* factor of *n* for which *n/f* is a factor of *m*.

Returning to our example, the smallest factors of 60 (*n*) are 1, 2, 3, 4, 5, 6, 10 ... (being the possible *f* values). The corresponding *g* vales are 60, 30, 20, 15, 12, 10, 6 ... Now the smallest *f* value such that the corresponding *g* value divides evenly into 264 is 5. (The *g* value here is 12, and 12 × 22 = 264, the *m* value.) Thus repetitions begin after the 5th iteration.

These considerations lend support to the view that when *n* is a prime number, the first repetitions occur only after *n* iterations, never earlier. For if *n* is prime, it has no factors other than 1 and itself. Our *f* value can't be 1, since the corresponding *g* value would be equal to *n* and have to divide equally into *m*. But by our explicit limiting condition, *n* is not a factor of *m*. So *f* can only be *n* and never smaller than *n*.

Bearing this in mind, we can simplify our conjecture as:

> For any two numbers m and n such that n is not a factor of m, the fractional part of m/n will first repeat after f iterations where f is the *smallest* factor of n for which n/f is a factor of m.

For where n is a prime number, n/f (that is, g) can only be 1, making f equal to n.

Alas, we have not got a watertight argument here. All that our dissection of n into factors f and g has shown is that the fractional part will repeat after f iterations, not that it will *first* repeat after f iterations. Better minds are needed to turn that conjecture into a proof.

A perfect difference

Do you recognise this sequence of numbers?

6, 28, 496, 8,128, 33,550,336, 8,589,869,056, ...

This is the beginning of the sequence of numbers known as *perfect numbers*. A perfect number is a positive integer that is equal to the sum of its proper positive factors. The proper positive factors of 6 — that is, the positive whole numbers that divide evenly into 6 — are 1, 2 and 3. And it just so happens that $1 + 2 + 3 = 6$. Likewise, the proper positive factors of 28 are 1, 2, 4, 7 and 14, and the sum of these factors is 28.

Let's finish with a puzzle. The string of digits below is made up by interweaving digits from two prime numbers.

21990297

The digits of each number are presented in the right order and the difference between the two numbers is a perfect number. What are the two prime numbers?

You don't need to consult the table of prime numbers in appendix B to solve this problem. (Indeed, that would make the exercise tedious and soporific.) Simple arithmetic with a dash of logic will give you the solution. You will only need to consult the appendix to confirm, if you need to, that certain numbers are indeed prime.

The answer is on page 189.

$$(n) \sim \frac{\sqrt{2n}}{\log n} \prod_{w=3}^{\infty} \left(1 - \frac{1}{w-1} \cdot \left(\frac{2n}{w}\right)\right)$$

Part 2

Appendices

A: The pitfalls in measuring intelligence

It is, perhaps, a platitude to say that intelligence is highly regarded in most societies. The wooden-headed are invariably scorned and often mocked. They are shepherded into special classes at school and, later in life, find themselves in dull jobs. Their opinions are mostly spurned — other than as the butt of jokes — and they tend to remain on the periphery of power and prestige. Better by far to be (or be seen to be) intelligent. So it is hardly surprising to read that:

> "scores on intelligence related tests matter, and the stakes can be high."[1]

Intelligence is one of society's most prominent discriminators. Age, fitness and gender are perhaps the only discriminators more prominent, at least in secular countries. There is some justification for age discrimination (at least at the lower end of the scale). However much a ten-year-old wants to work in a mine rather than study at school, there are good reasons why society is justified in forbidding that. The discrimination against an amputee wanting to work as an airline pilot also appears justified, though obviously for different reasons. Discrimination based on gender is not justified under any circumstances (a fact not annulled by our seeming inability to eradicate it).

---

1.   U. Neisser et al., "Intelligence: Knowns and Unknowns", *American Psychologist*, vol. 51, no. 2, 1996, p. 78.

What of intelligence? It would seem unfair to deny someone a job where intellectual prowess is not needed solely on the grounds that their score on some test was to the left of the bell curve of general intelligence. But just as some jobs require limbs, others require a sharpness of mind not possessed by everyone. A neurosurgeon who encounters an unexpected bleed while debulking a tumour might need to make a split-second decision if the patient is to be saved, as might a pilot encountering an unexpected intrusion into allocated airspace. A society as complex, and expectant, as ours needs quick-thinking highly intelligent members. Our welfare depends on it.

Society tends to reward the more intelligent more than the less intelligent (although that is not always the case: a dim-witted footballer can earn millions a year while a sharp-minded scientist might earn a fraction as much). The reasons are not always obvious or straightforward. But whatever they are, rewards are finite and the demand for them is invariably greater than the size of the pot. So some form of discrimination is necessary. A country of 20 million people cannot afford to have a million neurosurgeons or airline pilots. The discriminant mostly used to select who gets rewarded and who misses out is a blend of knowledge and intelligence (although some professions will call for other attributes as well). And in that blend, intelligence is usually given precedence. Ten trainee pilots applying for the one job might have had exactly the same training—and thus the same knowledge—but only one can be chosen. One hundred students might have been taught exactly the same things and thus have exactly the same knowledge, but if they are all applying for a single scholarship, then obviously something more than just knowledge is needed as the discriminant. In such cases, the discriminant society has chosen is usually intelligence, and our degree of intelligence is usually computed from the results of an intelligence test.

> "testing of abilities has always been intended as an impartial way ... of determining who gets what [such as] power and privilege, responsibility and reward."[2]

> "In the United States today, high test scores and grades are prerequisites for entry into many careers and professions."[3]

Since our personal future could be determined in some not inconsiderable way by the score we are given on an intelligence test, it is critical that intelligence tests really do measure the thing that society expects them to measure: intelligence. If someone is to be denied power, privilege, responsibility or reward on the basis of a test, the test had better be a good one.

But what is intelligence? Alas, it seems that not even psychologists agree on what it is:

> "when two dozen prominent theorists were recently asked to define intelligence, they gave two dozen somewhat different definitions"[4]

If there is no strong agreement on what intelligence is, how can we be confident that any particular intelligence test is actually measuring intelligence? Perhaps the tests are good at predicting achievement at school—although that correlation is only 0.5[5]—but does that warrant them being called tests of *intelligence* (with all the highly charged connotation that that word carries)?

That doubt can be magnified by a close examination of some of the questions asked on standard intelligence

---

2. L. J. Cronbach, *Essentials of Psychological Testing*, Harper & Row, New York, 1984, p. 5.
3. U. Neisser op. cit., p. 96.
4. ibid., p. 77.
5. ibid., p. 96.

tests. It seems in keeping with the common-or-garden idea of intelligence that the number of correct answers given on an intelligence test would be the prime measure of intelligence: the more correct answers, the more intelligent. That is no doubt the assumption of those who are being tested. But it is not the case, as I'm about to show.

Psychologists will claim that it is "a weakness [if a question on an intelligence test] allows the putting of an ingenious alternative answer by some high-grade subjects".[6] Yet many intelligence tests pose questions for which there is not one correct answer. Indeed, a testee doesn't have to be especially "high-grade" to see patterns in number sequences and visual sequences that the setter of the test has not seen. We explored this with some examples earlier in this book. On page 27 we showed that you don't need an Einsteinian brain to see that the IQ test question:

3. Please enter the missing figure: 4, 5, 8, 17, 44,

A. 80
B. 125
C. 112
D. 60
E. 84

admits to more than one of the given answers. Likewise with this IQ test question explored on page 46:

2. Find the picture that follows logically from the diagrams to the right.

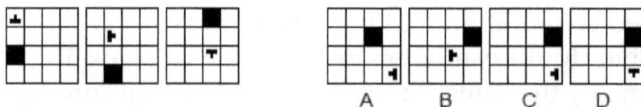

Unless blinded by hubris, creators of IQ tests must, surely, carry the nagging doubt that creative minds might well see a pattern that did not occur to them.

6.   E. Anstey, *Psychological Tests*, Thomas Nelson, London, 1966, p. 162.

Let's return to the type of IQ test question that mostly occupied us earlier in this book: the number-sequence question. A very simple example is {2, 4, 6, 8, 10, ?}. As I showed earlier, any number can continue the above sequence—can continue any finite sequence—and the sequence so continued will be the product of a strict mathematical rule (an equation, in other words). So there will be an infinite number of such rules capable of generating any given sequence. In other words, there is an infinite number of solutions to any number-sequence problem. So, a "high-grade subject" knowing this to be the case, might give the correct answer— namely, "any number"—and be marked wrong. Their answer will be marked wrong if they are responding to a multiple-choice question (where the assumption is explicit that there is only one correct answer and it is a number). It will be marked wrong if the marking is done not by a human but by an optical scanner. And it will be marked wrong if marked by a human who is merely following the answers provided by the test designers. In other words, the correct answer is incorrect. Is that intelligent?

Psychologists have long considered number-sequence problems to be particularly useful in intelligence testing, some even suggesting that such problems "provide one of the best tests of reasoning".[7] Not all intelligence tests include such problems, but many do.[8] But if they admit to an infinite number of answers—equivalent to one correct answer: "any number"—how useful can they be?

---

7. P. E. Vernon, *Intelligence and Attainment Tests*, University of London Press, London, 1960, p. 81.
8. Some examples: the AQ, MQ and PQ tests developed by the Australian Council for Educational Research. In fact, 14 of the 34 questions on the PQ test are of the standard number-sequence variety, with another 9 being a slight variation on the theme.

§

When the ambiguity of a question on an intelligence test is pointed out, three defences are often put forward:

- The answer being sought is the logical or most logical of all the possible answers.
- The answer being sought is the simplest or most obvious of all the possible answers.
- It doesn't matter what answer is given so long as it is the answer given by most people who have been independently verified as highly intelligent.

The first defence can be dispensed with easily. We considered earlier an example of an IQ test question that asked for the *logical* answer, namely:

2. Find the picture that follows logically from the diagrams to the right.

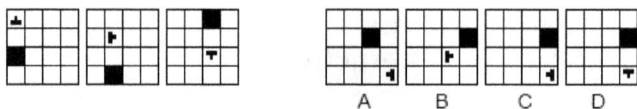

The answer the setter of the puzzle gives is C, so the generating algorithm must be considered to be logical, for presumably if y logically follows from x according to algorithm z, algorithm z must be logical. We suggested earlier that answers A, B and D are also correct (see page 27) and invited you to construct algorithms that will generate them. (One set of algorithms is given on page 187.) But we now should consider whether A, B and D "follow logically" from the input pictures given. In other words, are the algorithms that generate A, B and D logical?

Such a question can be easily dismissed because it is tautologous. It is equivalent to asking "is red a colour?". For surely the only way that an algorithm could be considered *logical* is if its internal logic or reasoning succeeds in generating the required output from the given input. We have shown that there are algorithms that do succeed in generating A, B and D from the given input, so are they not also logical? Likewise, we earlier produced polynomials that generate sequence continuations that test-setters would mark as wrong despite those continuations being unmistakably logical. They are, clearly, the result of applying iterative rules just as is done in deriving, say, pattern A.

Perhaps the question-setter meant that we are to find the picture that follows *most* logically. But it makes little sense to ask if one algorithm is more logical than another. If one computer programmer uses, say, the Miller–Rabin primality test to determine if a number is prime and another programmer uses the Baillie–PSW primality test, it would be bizarre to call one programmer's algorithm more logical than the other. Perhaps one algorithm is more *efficient* than the other—it might take less processing time—but greatest efficiency is not the same as most logical. (Is a company that can create 100 widgets with 10 staff less logical than a company that can make 112 widgets with 9 staff?) *Logical* is an absolute adjective. It has no comparative or superlative forms.

We are left to wonder whether what the question-setter meant by "follow logically" is "follow logically and most efficiently". If so, the question should have asked that. It is, surely, unfair to mark an answer as wrong if the answer was meant to be got in a particular way and that way is kept hidden from the answerer. Anyway, the algorithms for B and D given on page 188 have fewer steps than the algorithm for C, the preferred answer. So on a *logical-plus-most-efficient* interpretation of "logical", the author of the question has given the wrong answer.

The second defence—that testers are looking for the simplest or most obvious answer—is the position of the late Victor Serebriakoff (co-founder of Mensa and great fan of intelligence tests), set out in personal correspondence. That intelligence tests never explicitly ask for the simplest or most obvious answers may be in recognition of the fact that the meaning of *simplicity* and *obviousness* is neither simple nor obvious.

Consider, firstly, the meaning of *simple*. In the context, say, of answering an arithmetical question on an intelligence test—for example, {0, 1, 3, 6, 10, ?}—there seems to be only two likely candidates: (a) simple in the sense of requiring few steps or (b) simple in the sense of being easy.

By the first definition, one answer to a question will be simpler than another if it was reached with fewer steps. But suppose we have two possible answers to an arithmetical question, both employing the same number of steps but one using more-advanced arithmetic. An example is {2, 4, 16, 128, ?}. One answer is 1240, derived by noting, firstly, that the numbers increase by {2, 12, 112, ...} and, secondly, that these differences increase by {10, 100, ...}, a set that could conceivably continue with 1000. So if x − 128 = 1112, then x = 1240. Another possible answer is 2048, derived by noting that the numbers in the sequence are all powers of 2, the indices form the set {1, 2, 4, 7, ...} and that the differences between consecutive indices forms the sequence {1, 2, 3, ...}, a sequence which could conceivably continue with 4. And 2^{11} is 2048. Note that both answers require two steps: in the first case we derive one set of differences and then another; in the second we derive indices and then the set of differences between those indices.

But are the two answers as simple as one another? The answer is surely no, for powers are certainly more complex than subtraction (which is why powers are taught later than subtraction at school). Hence the first answer in our example (1240) would have to be considered simpler than the second even though it needed just as many steps.

Thus treating simplicity as a measure of the number of steps involved in reaching an answer is inadequate. We need also to be able to rank the various types of arithmetical operations in terms of some other sort of simplicity, and the most likely candidate appears to be simplicity as easiness.

But now the waters cloud with a good deal of subjective pollution. Obviously powers should, on our new definition, be ranked as less simple than subtraction. But is multiplication less simple than addition? Perhaps it is simpler than division, but is it simpler than powers? (Aren't powers just a form of multiplication?) And doesn't it all depend on the size of the numbers being manipulated? For example, 23,456,876 plus 213,546,765 might not be as simple as 2 times 3, but it is simpler than 234,321 times 216,876.

It's now getting messy and complicated and, to make matters worse, the ranking derived will also need to be weighted. For if, say, addition is simpler than powers, might it nevertheless be the case that an answer requiring two steps of addition is less simple than an answer requiring just one step of powers? Similarly, might not two steps of addition each involving three-digit numbers be less complex than three steps of addition each involving two-digit numbers?

This suggests that the arithmetical simplicity of an answer is a compound of simplicity in both senses outlined earlier. In other words, it is a function both of the number of steps taken to reach the answer and the relative easiness of the operation used in each step. For the sake of an illustration, let's say that single-digit addition is given an easiness weighting of 1.0, two-digit addition a weighting of 1.2 and powers to base 2 a weighting of 2.3. It would follow from these weightings that an answer involving one step of powers to base 2 is less simple than an answer involving a step of single-digit addition and a step of two-digit addition.

But no mathematician has established — or is ever likely to establish — a table of objective rankings and weightings for all the arithmetical operations (taking into account the various degrees of complexity due to the various sizes numbers can take). The task would be daunting and the objectivity doubtful. But without this information, how can we ever prove that the x steps used in deriving answer p indicate an answer simpler than q derived using y steps?

Even if some objective way of quantifying this concept could be found, we would still need to radically rethink the way intelligence tests are administered and marked. For if simplicity is partly a function of the number of steps needed to derive an answer, a criterion that links correctness to the simplest answer means that testers need to see not just the answer to a question but also the method the testee used to derive it. For if two testees arrived at the same answer but one took more steps, then the answer derived using more steps would need to be marked as incorrect even though it is identical to the answer offered by the other testee. That testees are not asked for their methods as well as their answers suggests that psychologists do not consider simplicity to be an important factor in determining the acceptability of answers to questions on intelligence tests.

As an example of the one sequence continuation that has numerous explanations, consider again the sequence discussed on page 8, namely {8, 16, 24, 32, ?, ? ...}. One solution is {8, 16, 24, 32, 40, 48 ...}. But there are at least five ways of explaining this particular continuation:

- Each number in the sequence is eight times its position in the sequence.
- The numbers represent the differences between the squares of consecutive odd digits.
- The numbers represent the midpoints between the differences between the squares of consecutive even digits.
- Take the differences between the products of consecutive natural numbers and multiply each by 4.
- Take the differences between the differences between consecutive cubes, beginning with zero, To each in turn add $2(n - 4)$, where n is the place of each term in the sequence.

We also saw that the answer IQ Labs gave to their puzzle {4, 5, 8, 17, 44, ?}—namely 125—could be reached in at least five ways (see page 27). That is, five quite different reasons could be given for answering 125. To ask markers of IQ test answers to adjudicate between very different reasons for continuing a sequence in a particular way is to impose on them an onerous responsibility, made more so by the necessary arbitrariness of their decision. Why is seeing 125 as following because the difference between the numbers follows the series {1, 3, 9, 27, 81 ...} any more (or less) intelligent—or more or less simple—than seeing it follow because each figure is three times the previous figure less 7?

Failing to ask testees for their reasons in arriving at their answer is potentially to deny the truly brilliant. But to ask testees for their reasons assumes that there is some objective way of ranking answers. Either way, the test setter is facing a Gordian knot impervious to even the sharpest blade.

Consider, now, the concept of *obviousness*. The popular British puzzlists Gyles Brandreth and Trevor Truran proposed this devilish number-sequence problem in one of their many books: {1, 8, 3, 7, 1, 9, 0, ?}. If you had a passion for British history, an obvious answer might be 1, the very answer the authors give (1837–1901 being the reign of Queen Victoria). That this would hardly be considered obvious by many others indicates, perhaps, that obviousness is an intractably relative concept. A trivial example, perhaps, so let's consider an example where numerical reasoning is expected: {3, 1, 4, 1, 5, ? ...}. Some might opt for 1; others for 9 (on the grounds that the digits are the leading digits of π). And another example: {4, 8, 32, 512, ?}. Once again the obvious answer is not so obvious. To some, the numbers in this sequence will be seen immediately as powers to base 2, with the indices forming the sequence {2, 3, 5, 9 ...}. The differences between these indices form the sequence {1, 2, 4 ...}. Is it obvious, now, how this new sequence continues? Is each number twice the previous number, leading to 8 as the continuation? Or is the sequence proceeding by progressive addition (+1, +2, +3 ...) leading to 7 as the continuation? It is certainly not obvious which of these continuations is the most obvious.

Of course, some would not see {4, 8, 32, 512 ...} as a set of powers to base 2. They might see each number as half of the square of the previous number. Others might think that adjacent pairs are being multiplied together and the product multiplied by the corresponding number in the sequence {1, 2, 3, 4 ...). Here, then, are four non-obscure ways of continuing {4, 8, 32, 512 ...} each of which could reasonably be considered by those who adopted them as obvious. But how are we to judge which is the *most* obvious?

In one sense it makes no sense to talk of something being *more* obvious than something else. Something is either obvious or it is not. The word *obvious* — like the word *logical* — is an absolute adjective: it has no comparative or superlative forms. In this sense, something is obvious if it is apparent or evident. If someone who has given 3 as the square root of 9 is told that –3 is also a square root of 9, –3 becomes an obvious answer. But it is not less obvious than 3 for not having been known at the time the question was asked. An appropriate parallel is this: answers to the question "where can I ski in Australia?" might include "Thredbo" and "Falls Creek". To someone who knows that you can ski at these places, the two answers are obvious. But one answer is not more obvious than another: they are just both obvious. Suppose, now, that someone answered this question with "Perisher". To someone who knew you can ski at Perisher, this answer is obvious; to someone who didn't, it is not. But once they know you can ski at Perisher, "Perisher" becomes an obvious answer. Further, it wasn't a less obvious answer beforehand; it just wasn't obvious at all. To return to sequences: two people might each see {4, 8, 32, 512 ...} as a set of powers to base 2 with the indices forming the sequence {2, 3, 5, 9 ...} and the differences between the indices forming the sequence {1, 2, 4 ...}. They might then interpret the latter sequence differently. One might give 128 as the answer (if the sequence of differences is interpreted as continuing with 7) and the other might answer 256 (if the sequence of differences is interpreted as continuing with 8). Each might think that their own answer is obvious. But when told the other's answer, each is more likely to say "That wasn't obvious!" rather than "That is less [or more] obvious than the answer I got!".

It might be retorted here that "Perisher" is a less obvious answer because it is not as widely known that you can ski at Perisher as it is that you can ski at, say, Thredbo. This introduces a different sense of *obvious*. In this sense, a more obvious answer is one that is more widely known, and the most obvious answer would be the one most people would give.

Applied to intelligence tests, this new sense of *obvious* suggests that test constructors should give a particular problem to a sample of people and declare the most obvious answer to be the one that most people in that sample arrived at. From then on, that answer will be the accepted answer.

But this approach is highly questionable. It is axiomatic that the highly gifted are disposed to give answers to test questions that the majority do not give. (That is partly what sets them apart.) Indeed, such people will often deliberately give a more obscure answer to an intelligence test question thinking that more marks might be awarded for being bright enough to detect an answer more complex but equally correct. To penalise the gifted for doing so is contrary to one of the purposes of intelligence tests: to identify those of superior intellect.

Anyway, it should be clear that much that is not obvious is nonetheless true. Quantum entanglement is not at all obvious — it still baffles physicists — and yet its existence is incontrovertible. To many minds, the Newtonian notion of gravity is more obvious — and more visualisable — than the Einsteinian notion, and yet the latter is doing a better job at explaining the universe. Once it was obvious that stress was the cause of stomach ulcers: that is, until a couple of scientists bravely swallowed a beaker of bugs and proved that the cause was the bacterium *Helicobacter pylori*. The power that obviousness has as a marker of intelligence is beginning to look rather thin indeed.

Anyway, would we always consider a person who saw the simplest or most obvious pattern in something to be more intelligent than the person who saw something more complex? Consider two poetry critics. One draws the reader's attention to the rhyming lines of a poem; the other to the fact that the lines are structured as iambic pentameters. The depth of observation of the second critic surely deserves more esteem than the simpler observation offered by the first. Or consider two art lovers commenting, let us say, on Géricault's *The Raft of the Medusa*, one of the finest paintings of French Romanticism. The painting is of a raft carrying the survivors from the *Méduse*, which ran aground in 1816. Only 15 of the 147 people who were set adrift on a hurriedly constructed raft survived, having endured starvation, dehydration and cannibalism. Suppose that our first art lover notes that the painting depicts a moment from the aftermath of a shipwreck — an observation of great simplicity and unmistakable obviousness. Perhaps they might add that some on the raft have noticed another vessel in the distance and are excitedly waving to it, while others — such as the elder draped in red cloth — are unmoved, forlorn beyond grief and hope. There is, clearly, a little more complexity in what has been added. But compare all that with this:

> "the response Géricault seeks is one beyond mere pity and indignation, though these emotions might be picked up *en route* like hitchhikers. For all its subject-matter, [the painting] is full of muscle and dynamism. The figures on the raft are like the waves: beneath them, yet also through them surges the energy of the ocean. Were they painted in lifelike exhaustion they would be mere dribbles of spume rather than formal conduits. For the eye is washed — not teased, not persuaded, but tide-tugged — up to the peak of the hailing figure, down to the trough of the despairing elder, across to the recumbent corpse front right who links and leaks into the real tides. It is because the figures are sturdy enough

to transmit such power that the canvas unlooses in us deeper, submarinous emotions, can shift us through currents of hope and despair, elation, panic and resignation."[9]

Two art lovers are looking for patterns in the same painting. One gives a relatively simple description of the main themes; the other a far richer and more complex description. In the absence of other considerations, few would doubt that the second description must have emanated from a mind more intelligent than the first. Thus it is not the case that noticing the simplest of facts and patterns is what sets the intelligent off from those who are less intelligent. Thus we have reason to doubt that the person who answers {4, 5, 8, 17, 44, ?} with 125 is necessarily more intelligent than the person who answers 112 (to refer back to the IQ test question we analysed on page 27). So might not a testee who understands that in many fields—such as poetry and painting—discerning a more complex structure is deemed more estimable deliberately opt for the more complex of two possible answers? And might not that testee be marked wrong despite having the intelligence to discern several patterns when another testee is marked right despite having only discerned one pattern, namely, the simplest? Thus the claim that simplicity and obviousness is important in judging intelligence is too simple to be useful.

The third defence given to allowing questions on intelligence tests to have more than one answer is that it doesn't matter what answer is given so long as it is the answer given by most people who have been independently verified as highly intelligent. This is the defence most commonly given by psychologists and

---

9. J. Barnes, *A History of the World in 10½ Chapters*, Jonathan Cape, London, 1989, pp. 136–137.

others in the cognitive testing profession (often called psychometricians). So if most of those deemed, by some independent test, to be highly intelligent give 125 as the answer to {4, 5, 8, 17, 44, ?} then 125 is the only acceptable answer. There are two problems with this defence: runaway regression and inconsistent relativism.

Consider a particular test, say, T_1. If T_1 is deemed to be a good proxy measure of some attribute (such as intelligence) because its scores correlate strongly with the scores on some other test for that attribute (say T_2), then it is legitimate to ask how is it that we can be confident that T_2 is a good measure of intelligence. To answer that we know that T_2 is a good measure of intelligence because its scores correlate strongly with the scores on some other test for that attribute (say T_3) merely invites the same question, pushed back one level in a potential infinite regress. What makes T_3 a valid measure? If T_4, what makes it a valid answer? And so and so on. The regress has to stop somewhere if this defence is to proceed. And it can only stop at some robust, uncontroversial test or measure of intelligence (perhaps something akin to an axiom or definition). Let's call it T_7. Now if T_7 is obviously a better test of intelligence than T_1, why not simply rely on T_7 and discard T_1? To retort that T_1 is easier or less costly than T_7 will be of little comfort to those whose education or career has been thwarted or mis-channelled courtesy of an inferior intelligence-testing regime.

While on the subject on inter-test correlations, it is puzzling why designers of intelligence tests are keen to note that their test correlates well with other tests. (This is usually referred to as a measure of *concurrent validity*.) The fact that test T_3 correlates well with T_2 and T_2 correlates well with the supposedly uncontroversial T_7 doesn't mean that T_3 correlates well with T_7. Correlation comparisons are noncommutative. Consider, for example, the following three datasets: A, B and C.

A	B	C
5	6	5
8	9	6
7	9	7
2	3	5
8	7	7
9	7	5
4	6	5
6	8	9
6	6	5
7	4	5

Imagine that A is a list of the results from a test deemed to the best measure of some attribute or other, and B and C are the results from proxy tests alleged to be good measures of the same attribute. The correlation between B and A, and between C and B, is approximately 0.6 in both cases. You might think then that the correlation between C and A should be about the same. But it is just 0.24.[10] So the fact that a test has a good correlation with, say, a Wechsler intelligence test tells us nothing about how well the test correlates with the test that gave the Wechsler test its validity (if it has validity).

Let's turn now to the second problem with basing correctness on the answers of those independently judged to be intelligent: relativism. Suppose that more and more people begin answering "any number" to number-sequence

---

10. You can easily check these correlation values using the CORREL function in Microsoft Excel.

questions on intelligence tests. This is to be expected given how well entrenched is the notion that intelligence is inextricably linked with an ability to provide *correct* solutions to problems. Psychologists will, presumably, observe this trend and seek to rediscover the answers intelligent people give to number-sequence questions (for it is only on this basis, they argue, that intelligence can be attributed or denied). It is very likely, then, that at some time psychologists will discover that more than the critical number of intelligent people—the majority, the top 40%, those in the top three percentiles, or however it is defined—are now answering "any number" rather than 125 when given {4, 5, 8, 17, 44, ?} on an intelligence test. Suppose that this critical number is reached at time t_2. Suppose further that testee A takes an intelligence test at t_1—which precedes t_2—and takes the same test again at t_3 (which follows t_2). Let us suppose also that t_1 and t_3 are not too far apart and that A is of an age at which intelligence is fairly stable. If A answers each question identically on both occasions—including answering the number-sequence problems with "any number"—then A is, presumably, more intelligent at t_3 than at t_1. This is indeed odd, but what is odder is that A's level of intelligence has changed solely because of a change in the answers *other people* have given to test items. This extraordinary, even absurd, conclusion follows from an insistence that a person is intelligent solely if they do what other allegedly intelligent people do.

There is a telling corollary to the argument from relativism we have just been considering. Are the truly intelligent those who think within the box or outside the box? Is not the *original* thinker more often than not considered more intelligent than those who just go along with the accepted view—even the accepted view of those *generally* considered to be intelligent? Nineteenth-century professors of physics—who no doubt would have been considered highly intelligent—all held a similar view of gravity (based on the work of Isaac Newton). Along comes Albert Einstein. His idea of gravity didn't match that of those deemed to be highly intelligent. But, clearly, Einstein was highly intelligent. Indeed, his idea of gravity has come to supplant the Newtonian idea. Do we want to encourage more Einsteins? Or just those who think like others (however we define their cohort). Or to put it into the earlier vocabulary of this book, is it more intelligent to see patterns that others see, or to see entirely new patterns?

Further, if we expand the *what-the-intelligent-person-says* argument to wider contexts than just IQ test questions, we have to ask whether it is always intelligent to believe what the intelligent believe. Suppose that 80% (or whatever figure you care to consider) of some cohort determined in some way or other to be intelligent believe that the universe was designed by a god. Is that a good reason to argue that it is intelligent to believe that that god exists? What if the majority of those deemed intelligent believe—as they once did—hat the sun revolves around the earth? Would it have been a sign of intelligence to go along with the crowd and believe the same? So just because the majority of successful and cultured folk[11] answer 125 to {4, 5, 8, 17, 44, ?} cannot mean that it is not intelligent to answer, say, 112, or 84, or any of the other answers we considered possible on page 27.

Psychological testing is a serious business. A person's future might be mapped out, and their self-esteem

---

11. Assuming that these are some of the characteristics that are considered important in independently determining a person's intelligence (that is, used to assemble a cohort that could be used to validate an intelligence test).

affected, by the result they achieve on such a test. Where a test item calls for a closed response—as in the case of a number-sequence question—testees rightfully expect to be rewarded for getting the answer correct, and this means that the only acceptable answer should be the correct answer. Furthermore, employers, schools and universities using tests to select only the ablest testees will be poorly served by tests that can rank those who give the correct answer below those who give the standard or expected answer. Neither of these desirable outcomes—rewarding correctness and selecting the ablest—is guaranteed by the use of number-sequence problems in such tests. Nor are they guaranteed by the comparative approach to intelligence testing (which, of course, exerts its levelling influence over all types of intelligence-test questions).

The solution in the case of number-sequence questions is not to revise them so that they might become better test items. That is not possible if, at the same time, such items are to retain their purpose as discriminators of numerical ability. For to revise such items so that the acceptable answer is always the correct answer means that every such item (whatever the numbers in the given sequence) would have the very same answer: "any number". And this answer, once known, ceases to be the product of numerical reasoning. The solution, then, can only be to discard these questions altogether from psychological tests.

§

Intelligence tests might be good at discriminating some attribute or other (such as likely school performance), but they also discriminate in another sense—the pejorative sense of implied unfairness. For a start, intelligence tests have led to some horrific classification errors. American psychologist Lee Cronbach—a fan of tests—observed that:

"At times unsound testing or bad interpretations of
tests have assigned children to the retarded category
who did not belong there."[12]

For a youngster to be labelled retarded—especially when
they are not—would be felt as nothing short of life-
shattering. For Cronbach, what redeemed the situation
was that:

"[at least] the majority of children [suspected of]
retardation [were] saved [by a further] psychological
examination".[13]

Well that's cold comfort for the approximately 60,000
Americans who were forcibly sterilised between 1907 and
1968 based in part on their scores on intelligence tests.[14]

Carrie Beck became famous in 1924 for challenging
the sterilisation laws, after having been declared retarded
and committed to an institution. She lost her case and was
forcibly sterilised, as was her sister Doris. Stephen Jay
Gould, writing about Carrie and Doris, states:

"Neither would be considered mentally deficient by
today's standards ... Can one measure the pain of a
single dream unfulfilled, the hope of a defenseless
woman snatched by public power in the name of an
ideology advanced to purify a race?"[15]

Eugenics is a type of discrimination that is clearly immoral.
Happily, it is largely a practice of the past. But discrimination based on intelligence test scores is still rife, even if
the outcomes are less horrific than forced sterilisation. No
intelligence test correlates perfectly with whatever gold

---

12. L. J. Cronbach op. cit., pp. 4–5.
13. ibid., p. 5.
14. P. R. Reilly, *The Surgical Solution: A History of Involuntary Sterilization in the
    United States*, The Johns Hopkins University Press, Baltimore, 1991, p. 94.
15. Stephen Jay Gould, *The Mismeasure of Man*, Norton, New York, 2nd ed.,
    1996 , p. 335.

standard it is being compared with. For example, if a test is being used to predict school performance, the correlation would be perfect if the correlation coefficient was +1. This would mean that everyone who scored highly on the test did very well at school and everyone who did badly on the test did badly at school. (It would also be perfect if the correlation was −1, but now a high score on the test would correctly predict poor school performance and vice versa.) But, as we've noted, the correlation coefficient is not perfect. It is about 0.5. This means that some who do well on the test will do poorly at school and some who did poorly on the test will do well at school. Consider the data in Table 20 below.

Table 20: What might a correlation of 0.5 signify?

Test	School	Test Mean	School Mean	Inconsistent
104	8	104.25	7.7	Yes
106	7			Yes
98	9	Correlation		Yes
87	8	0.50051894		Yes
110	9			
120	8			
115	9			
104	8			Yes
111	9			
90	6			
102	9			Yes
110	10			
125	8			
118	9			
106	7			Yes
90	4			
105	7			Yes
110	5			Yes
99	9			Yes
75	5			

In the first column are the scores of 20 students who sat an intelligence test and in the second column is how well each subsequently performed at school (on a scale of 1 to 10). The mean of each set of scores is given at the top of the table. I have concocted these scores, but that does not matter in the argument that follows. What does matter is that the correlation between the two sets of scores is 0.5 — the very same as is reported in the literature.

Now compare each score against the mean for that particular set of scores. You will notice that in six cases a test score less than the mean corresponds to a school performance score that is greater than the mean, and in four cases a test score greater than the mean corresponds to a school performance score that is less than the mean. In other words, in 50% of cases — those shown with **Yes** in the **Inconsistent** column — the intelligence test was a poor predictor of school performance. You might think this is only academic. But what if the test, being considered useful in predicting school performance, is used to distribute limited advantages (such as scholarships)? Is it fair that six out of the 20 students (30%) missed out on a scholarship despite having above average-academic potential? And is it an efficient use of scarce resources to give four students (20%) a scholarship when its value will only be squandered? Correlation is simply too blunt a measure to use in making such life-changing decisions.

The use of intelligence tests to gauge school performance is scarily self-validating. If a child doesn't score well on such a test, it is likely that they will feel inferior to those peers who did better. Perhaps they will be put in a stream for the less-smart and get to learn less. Perhaps they will suffer class snobbery and maybe even bullying. Might all this be enough to dull their motivation? The overall outcome is poor school performance — exactly as the test predicted. But what if they hadn't been given the test and had received the same encouragement and education as their peers? It is not conceivable that their school performance would have been better?

So intelligence tests can adversely discriminate against a minority (namely those who are bright but do not score well on the tests). But they can also discriminate against the majority. Students can be coached to do better at intelligence tests. A spokesperson for EduTest—an educational assessment organisation in Australia that designs intelligence tests for use in scholarship filtering—is reported as saying "We are constantly editing tests ... to stay ahead of coaching colleges".[16] This is an admission that you can coach a student to a better mark on a scholarship test. (Otherwise why bother trying to stay ahead of the coaches?) Hence parental wealth—and a willingness to provide your children with the best opportunities available—will in part influence who does well on an intelligence test and hence who gets a scholarship. Poor parents won't be able to afford coaches. Their children are thus less likely to get a scholarship. Awarding scholarships under a scheme whereby the children of the wealthy are likely to score higher than the children of the poor is a form of discrimination. In this case, it may be discrimination against the majority, for in many countries, the majority of families struggle just to maintain a decent standard of living. They do not have the disposable income necessary to pay for extra-classroom coaching for their children.

That coaching—and schooling in general—can affect IQ has been known for a while:

---

16. K. Marshall, "Fierce Contest for Edge in Education", *The Age*, 17 March 2014, p. 9.

"A striking demonstration of this effect appeared when schools in one Virginia county closed for several years in the 1960s to avoid integration [i.e., government-forced racial desegregation], leaving most black children with no formal education at all. Compared to controls, the intelligence-test scores of these children dropped by about 0.4 standard deviations (6 points) per missed year of school."[17]

If the removal of schooling can lower one's intelligence, it seems reasonable to assume that additional schooling (through coaching) is going to yield a higher intelligence score than might otherwise be the case. And experiments have indeed shown that coaching can give one a higher score on an intelligence test: up to 6 IQ points with coaching alone and 11 IQ points where the coaching includes practice tests.[18]

Discrimination, whatever its form, warrants attention. Some discrimination, as we noted earlier, is unavoidable. A one-armed person with no mind-controllable prosthesis should not have the same opportunities as an able-bodied person to become a commercial pilot. But here the reason for the discrimination is relevant. You need two arms to fly a plane. But a poor background does not limit one's ability to be pilot, a judge, a surgeon — indeed, any professional activity. It is simply not a relevant consideration. To limit one's access to a profession — and the rewards that go with it — on a basis of a tool that favours the wealthy is morally questionable, to say the least.

---

17. U. Neisser op. cit., p. 87.
18. R. L. Bangert-Drowns, J. A. Kulik & C. C. Kulik, "Synthesis of Research on the Effects of Coaching for Aptitude and Admission Tests", *Educational Leadership*, vol. 41, iss. 4, December 1983/January 1984, pp. 80–82.

§

So what do intelligence tests measure if, as seems to be the case, they don't measure intelligence? Perhaps at best they measure nothing more than a person's ability to answer the type of questions that typically appear on intelligence tests. That is meant to be a criticism of intelligence tests. But some psychologists have gone so far as to define intelligence as what these tests measure: "If we agree to define intelligence as what the tests of intelligence tests, there is a good deal that we can say about it."[19]

Now definitions, obviously, are essential in any advance of knowledge. But do we always advance our knowledge by redefining concepts in ways that don't match common usage? It is true that sometimes common definitions do need to be rethought. Our concept of, say, *consciousness* needed redefining so as to exorcise it of the unscientific concept of *soul*. The concept of a *proton* as a fundamental building block of matter also had to be redefined as we discovered it to be a composite particle: a collection of quarks. And there are plenty more examples like that. But we do not advance knowledge by redefining concepts so as to make them unrecognisable as the things that spurred our curiosity in the first place. If we are curious about how birds fly, we do not redefine the word *bird* to mean aeroplane and divert the discussion to aeroplanes. We want to know about birds, and we know that birds are simply not aeroplanes (unless lexicographers have been misleading us). Likewise, I know that someone who answers a number-sequence problem with "any number" is more intelligent than someone who gives a single number. That is encapsulated in what we understand by intelligence. Define it in some other way if

---

19. E. G. Boring, "Intelligence as the Tests Test It", *New Republic*, iss. 36, 1923, p. 37.

you like—test prowess, perhaps—but now we are not talking about the same thing. You have moved the discussion to other grounds. You have engaged in definitional fiat. Fine. Your argument might have some academic or armchair interest. But I still want to know about intelligence.

The standard intelligence test is primarily a test of one's numeric, verbal, logic and pattern-recognition ability. In addition to number-sequence and shape-sequence questions of the sort we've explored earlier, the tests ask questions like "Languages are to meaning as philology is to [erudition, philosophy, ethics, semantics or grammar]" and "At the end of a banquet 10 people shake hands with each other. How many handshakes will there be in total?"[20] It is not difficult to imagine that a highly gifted composer such as Stravinsky, a brilliant artist such as Monet or a revolutionary choreographer such as Balanchine would do poorly with such questions. Their minds are occupied with matters far less trivial than mind puzzles (and probably were so too during school years). And yet if intelligence is what is measured by intelligence tests, the likes of Stravinsky, Monet and Balanchine would probably be considered duffers. Even Einstein was considered a bit of a duffer at school. The absurdity that so many geniuses would have been considered duffers is no doubt part of the reason for the widespread interest in Howard Gardner's theory of multiple intelligences. Gardner extends the notion of intelligence to include musicality, kinaesthetic ability, interpersonal skills and self-reflective aptitude to the range of skills tested by the standard intelligence test.[21]

---

20. These questions were taken from a sample intelligence test published by IQ Test Labs. See http://www.intelligencetest.com/index.htm. Viewed 10 March 2014).

21. M. Gardner, *Frames of Mind: The Theory of Multiple Intelligences*, Basic Books, Philadelphia, 1983.

Such a view seems to accord more with our common understanding of intelligence. Paper-and-pen tests of intelligence simply can't measure these abilities, nor other abilities recognised as markers of intelligence, such as "giving an extemporaneous talk ... or being able to find one's way in a new town." (Neisser op. cit., p. 790)

§

To ensure that IQ tests do measure what they purport to measure, we need intelligent people to design them. But to determine the intelligence of a prospective test designer, we need a good intelligence test. It is a catch-22 situation, one that spreadsheet users might note as resembling that frustrating error message:

"Circular references unresolved".

B: The first 1000 primes

2	3	5	7	11	13	17	19	23	29
31	37	41	43	47	53	59	61	67	71
73	79	83	89	97	101	103	107	109	113
127	131	137	139	149	151	157	163	167	173
179	181	191	193	197	199	211	223	227	229
233	239	241	251	257	263	269	271	277	281
283	293	307	311	313	317	331	337	347	349
353	359	367	373	379	383	389	397	401	409
419	421	431	433	439	443	449	457	461	463
467	479	487	491	499	503	509	521	523	541
547	557	563	569	571	577	587	593	599	601
607	613	617	619	631	641	643	647	653	659
661	673	677	683	691	701	709	719	727	733
739	743	751	757	761	769	773	787	797	809
811	821	823	827	829	839	853	857	859	863
877	881	883	887	907	911	919	929	937	941
947	953	967	971	977	983	991	997	1009	1013
1019	1021	1031	1033	1039	1049	1051	1061	1063	1069
1087	1091	1093	1097	1103	1109	1117	1123	1129	1151
1153	1163	1171	1181	1187	1193	1201	1213	1217	1223
1229	1231	1237	1249	1259	1277	1279	1283	1289	1291
1297	1301	1303	1307	1319	1321	1327	1361	1367	1373
1381	1399	1409	1423	1427	1429	1433	1439	1447	1451
1453	1459	1471	1481	1483	1487	1489	1493	1499	1511
1523	1531	1543	1549	1553	1559	1567	1571	1579	1583
1597	1601	1607	1609	1613	1619	1621	1627	1637	1657
1663	1667	1669	1693	1697	1699	1709	1721	1723	1733
1741	1747	1753	1759	1777	1783	1787	1789	1801	1811
1823	1831	1847	1861	1867	1871	1873	1877	1879	1889
1901	1907	1913	1931	1933	1949	1951	1973	1979	1987
1993	1997	1999	2003	2011	2017	2027	2029	2039	2053
2063	2069	2081	2083	2087	2089	2099	2111	2113	2129

2131	2137	2141	2143	2153	2161	2179	2203	2207	2213
2221	2237	2239	2243	2251	2267	2269	2273	2281	2287
2293	2297	2309	2311	2333	2339	2341	2347	2351	2357
2371	2377	2381	2383	2389	2393	2399	2411	2417	2423
2437	2441	2447	2459	2467	2473	2477	2503	2521	2531
2539	2543	2549	2551	2557	2579	2591	2593	2609	2617
2621	2633	2647	2657	2659	2663	2671	2677	2683	2687
2689	2693	2699	2707	2711	2713	2719	2729	2731	2741
2749	2753	2767	2777	2789	2791	2797	2801	2803	2819
2833	2837	2843	2851	2857	2861	2879	2887	2897	2903
2909	2917	2927	2939	2953	2957	2963	2969	2971	2999
3001	3011	3019	3023	3037	3041	3049	3061	3067	3079
3083	3089	3109	3119	3121	3137	3163	3167	3169	3181
3187	3191	3203	3209	3217	3221	3229	3251	3253	3257
3259	3271	3299	3301	3307	3313	3319	3323	3329	3331
3343	3347	3359	3361	3371	3373	3389	3391	3407	3413
3433	3449	3457	3461	3463	3467	3469	3491	3499	3511
3517	3527	3529	3533	3539	3541	3547	3557	3559	3571
3581	3583	3593	3607	3613	3617	3623	3631	3637	3643
3659	3671	3673	3677	3691	3697	3701	3709	3719	3727
3733	3739	3761	3767	3769	3779	3793	3797	3803	3821
3823	3833	3847	3851	3853	3863	3877	3881	3889	3907
3911	3917	3919	3923	3929	3931	3943	3947	3967	3989
4001	4003	4007	4013	4019	4021	4027	4049	4051	4057
4073	4079	4091	4093	4099	4111	4127	4129	4133	4139
4153	4157	4159	4177	4201	4211	4217	4219	4229	4231
4241	4243	4253	4259	4261	4271	4273	4283	4289	4297
4327	4337	4339	4349	4357	4363	4373	4391	4397	4409
4421	4423	4441	4447	4451	4457	4463	4481	4483	4493
4507	4513	4517	4519	4523	4547	4549	4561	4567	4583
4591	4597	4603	4621	4637	4639	4643	4649	4651	4657
4663	4673	4679	4691	4703	4721	4723	4729	4733	4751
4759	4783	4787	4789	4793	4799	4801	4813	4817	4831
4861	4871	4877	4889	4903	4909	4919	4931	4933	4937
4943	4951	4957	4967	4969	4973	4987	4993	4999	5003
5009	5011	5021	5023	5039	5051	5059	5077	5081	5087
5099	5101	5107	5113	5119	5147	5153	5167	5171	5179
5189	5197	5209	5227	5231	5233	5237	5261	5273	5279
5281	5297	5303	5309	5323	5333	5347	5351	5381	5387
5393	5399	5407	5413	5417	5419	5431	5437	5441	5443
5449	5471	5477	5479	5483	5501	5503	5507	5519	5521

5527	5531	5557	5563	5569	5573	5581	5591	5623	5639
5641	5647	5651	5653	5657	5659	5669	5683	5689	5693
5701	5711	5717	5737	5741	5743	5749	5779	5783	5791
5801	5807	5813	5821	5827	5839	5843	5849	5851	5857
5861	5867	5869	5879	5881	5897	5903	5923	5927	5939
5953	5981	5987	6007	6011	6029	6037	6043	6047	6053
6067	6073	6079	6089	6091	6101	6113	6121	6131	6133
6143	6151	6163	6173	6197	6199	6203	6211	6217	6221
6229	6247	6257	6263	6269	6271	6277	6287	6299	6301
6311	6317	6323	6329	6337	6343	6353	6359	6361	6367
6373	6379	6389	6397	6421	6427	6449	6451	6469	6473
6481	6491	6521	6529	6547	6551	6553	6563	6569	6571
6577	6581	6599	6607	6619	6637	6653	6659	6661	6673
6679	6689	6691	6701	6703	6709	6719	6733	6737	6761
6763	6779	6781	6791	6793	6803	6823	6827	6829	6833
6841	6857	6863	6869	6871	6883	6899	6907	6911	6917
6947	6949	6959	6961	6967	6971	6977	6983	6991	6997
7001	7013	7019	7027	7039	7043	7057	7069	7079	7103
7109	7121	7127	7129	7151	7159	7177	7187	7193	7207
7211	7213	7219	7229	7237	7243	7247	7253	7283	7297
7307	7309	7321	7331	7333	7349	7351	7369	7393	7411
7417	7433	7451	7457	7459	7477	7481	7487	7489	7499
7507	7517	7523	7529	7537	7541	7547	7549	7559	7561
7573	7577	7583	7589	7591	7603	7607	7621	7639	7643
7649	7669	7673	7681	7687	7691	7699	7703	7717	7723
7727	7741	7753	7757	7759	7789	7793	7817	7823	7829
7841	7853	7867	7873	7877	7879	7883	7901	7907	7919
7927	7933	7937	7949	7951	7963	7993	8009	8011	8017
8039	8053	8059	8069	8081	8087	8089	8093	8101	8111
8117	8123	8147	8161	8167	8171	8179	8191	8209	8219
8221	8231	8233	8237	8243	8263	8269	8273	8287	8291
8293	8297	8311	8317	8329	8353	8363	8369	8377	8387
8389	8419	8423	8429	8431	8443	8447	8461	8467	8501
8513	8521	8527	8537	8539	8543	8563	8573	8581	8597
8599	8609	8623	8627	8629	8641	8647	8663	8669	8677
8681	8689	8693	8699	8707	8713	8719	8731	8737	8741
8747	8753	8761	8779	8783	8803	8807	8819	8821	8831
8837	8839	8849	8861	8863	8867	8887	8893	8923	8929
8933	8941	8951	8963	8969	8971	8999	9001	9007	9011
9013	9029	9041	9043	9049	9059	9067	9091	9103	9109
9127	9133	9137	9151	9157	9161	9173	9181	9187	9199

9203	9209	9221	9227	9239	9241	9257	9277	9281	9283
9293	9311	9319	9323	9337	9341	9343	9349	9371	9377
9391	9397	9403	9413	9419	9421	9431	9433	9437	9439
9461	9463	9467	9473	9479	9491	9497	9511	9521	9533
9539	9547	9551	9587	9601	9613	9619	9623	9629	9631
9643	9649	9661	9677	9679	9689	9697	9719	9721	9733
9739	9743	9749	9767	9769	9781	9787	9791	9803	9811
9817	9829	9833	9839	9851	9857	9859	9871	9883	9887
9901	9907	9923	9929	9931	9941	9949	9967	9973	

C: Useful functions in Microsoft Excel

Calculating digital sums

Some of our doodlings explore the digital sum of a number. This is just the sum of all the number's digits. Thus the digital sum of 397 = 3 + 7 + 9 = 19. There are a number of ways to calculate the digital sum of a number in Microsoft Excel, some easy, some not so. Perhaps the easiest way is to use Visual Basic for Applications (VBA) to create a function specifically to do the task. Here is the process:

1. Save your Microsoft Excel spreadsheet as a macro-enabled workbook. (The file extension will be .xlsm.)
2. Press ALT + F11. This opens the Visual Basic Editor.
3. From the project browser panel at the left of the window, click once on the name of the sheet or workbook in which you want to use your function.
4. From the **Insert** menu, choose **Module**.
5. In the editing panel that appears, enter the seven lines of code you see in the illustration at the top of page 176.

```
digital substraction.xlsm - Module1 (Code)
(General)                          ▼   SumDigits
Function SumDigits(ByVal Number As Long) As Long
    Number = Abs(Number)
    Do While Number >= 1
        SumDigits = SumDigits + Number Mod 10
        Number = Int(Number / 10)
    Loop
End Function
```

You can call your function anything you like, but make sure that the name you use — SumDigits in the example above — is identical on both the first and third lines of the code.

6. Press ALT + Q to leave the editor and return to your spreadsheet.

7. You can now type (or select) your function just as you type or select any in-built function in Microsoft Excel:

	C1	▼	f_x	=SumDigits(A1)	
	A	B	C	D	E
1	25145		17		
2					

Some doodlings explore what we called the digital root of a number. The digital root of a number is the sum of its digits, with the digits of the sum also summed until a single number is reached. Thus the digital root of 25145 is 8 (being $2 + 5 + 1 + 4 + 5 = 17 = 1 + 7 = 8$). In these cases, the custom function described above will stop after the first summation, giving you just the digital sum, and not necessarily the digital root. You could get the digital root by creating a VBA function—following the steps given above—but with the following code:

```
Book1 - Module1 (Code)

(General)                          ▼    SumDigits2                         ▼

        Do While Number >= 1
            SumDigits2 = SumDigits2 + Number Mod 10
            Number = Int(Number / 10)
        Loop
        SumDigits2 = SumDigits2 Mod 9
        If SumDigits2 = 0 Then
            SumDigits2 = SumDigits2 + 9
        End If
    End Function
```

But there is a much simpler solution, one that relies on a curious feature of arithmetic, namely, the remainder after dividing a positive number by 9 is the digital root of that number. Some examples: 248/9 is 27 with a remainder of 5, and 5 is the digital root of 248. And 9879/9 is 1097 with a reminder of 6, and 6 is the digital root of 9879. In mathematics we say 248 *modulo* 9 is 5, and 9879 *modulo* 9 is 6. And Microsoft Excel has an in-built modulo function, named MOD. Thus if you entered the formula =MOD(248,9), the answer returned is 5.

There is a complication we must consider. If a number, *n*, is perfectly divisible by 9—that is, there is no remainder—n mod 9 yields 0 and yet the digital root will by 9. So we need to enter an IF function in our formula to be sure of always getting the digital root. Assuming that the number whose digital root you want to find is in cell A1, your formula should be this:

```
=IF(MOD(A1,9)=0,9,MOD(A1,9))
```

There is one further complication to consider: modulo 9 of a *negative* number, n, is not the digital root of n; however, modulo 9 of the absolute value of n is. Thus our formula needs to be:

```
=IF(MOD(ABS(A1),9)=0,9,MOD(ABS(A1),9))
```

D1			f_x	=IF(MOD(ABS(A1),9)=0,9,MOD(ABS(A1),9))			
A	B	C	D	E	F	G	H
785			2				
-111			3				
81			9				

Reversing digits

A number of the doodlings in this book explore the arithmetic of reverse numbers. For example, in "Mirror equivalence" on page 96 we considered the result of squaring a number, reversing the digits of the square and then looking for integer roots of the reversed number. Again, there are a number of ways of reversing a number in Microsoft Excel, some easy, some not so. Perhaps the easiest way is to create a function that will reverse a number in one step. Here's the way to do it:

1. Save your Microsoft Excel spreadsheet as a macro-enabled workbook (with a .xlsm extension).
2. Press ALT + F11. This opens the Visual Basic Editor.
3. From the project browser panel at the left of the window, click once on the name of the sheet or workbook in which you want to use your function.
4. From the **Insert** menu, choose **Module**.
5. In the editing panel that appears, enter the three lines of code you see in the illustration at the top of page 179.

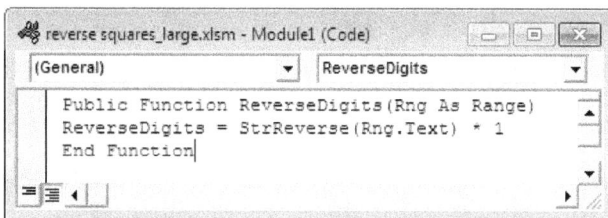

You can name your function anything you like, but the name you use—`ReverseDigits` in the example above—must be identical on both the first and second lines of the code.

6. Press ALT + Q to leave the editor and return to your spreadsheet.

7. You can now type (or select) your function just as you type or select any in-built function in Microsoft Excel:

Checking if a number is prime

Many of our doodlings explore prime numbers. There are array formulas you can enter in Microsoft Excel that will check if a number is prime. They are useful—if you can stomach their complexity. A simpler option is to create a user-function. Here is how to create one.

1. Save your Microsoft Excel spreadsheet as a macro-enabled workbook. (The file extension will be .xlsm.)

2. Press ALT + F11. This opens the Visual Basic Editor.

3. From the project browser panel at the left of the window, click once on the name of the sheet or workbook in which you want to use your function.

4. From the **Insert** menu, choose **Module**.

5. In the editing panel that appears, enter the lines of code you see in the illustration below.

```
IsPrime.xlsm - Module1 (Code)

(General)                                ▼   IsPrime                          ▼

Function IsPrime(Numb As Single) As Boolean
    Dim X As Long
    If Numb < 2 Or (Numb <> 2 And Numb Mod 2 = 0) _
    Or Numb <> Int(Numb) Then Exit Function
    For X = 3 To Sqr(Numb)
        If Numb Mod X = 0 Then Exit Function
    Next
    IsPrime = True
End Function
```

6. Press ALT + Q.

7. You can now type (or select) your function just as you type or select any in-built function in Microsoft Excel:

C1		fx	=IsPrime(A1)	
A	B	C	D	E
113		TRUE		

Note what the code does:

- The function aborts if the input number is less than 2 (since no prime number is less than 2).

- The function aborts if the input number is not equal to 2 and the remainder when the number is divided by 2 is zero. This gets rid of all the even numbers, since they are all evenly divisible by 2 (that is, modulo 2 = 0).

- The function aborts if the input number is not equal to 2 and it is not an integer (since all prime numbers are integers).

- The function now checks to see if there is a zero remainder when the input number is divided by any number between 3 and the square root on the input number. If, for any such number, there is no remainder, then the input number is divisible by another number and thus cannot be a prime.

- Note that the divisors to be checked need only go as high as the square root of the input number. For if a divisor is greater than the input number, then its partner—the number when multiplied by the divisor gives the input number—must be less than the square root of the input number. And any divisor less than that square root will have been checked in the iterations between 3 and that square root.

- We start the *For–Next* iteration with 3 because an input number if 1 would have aborted the function, and an input number of 2 would skip the iteration and generate a TRUE result (since 2 is a prime number).

D: From the SRSR mincer

In "Mirror equivalence" on page 96, I pondered the frequency and distribution of integers whose squares when reversed — that is, read from right to left instead of from left to right — become the squares of the reverse of the original number. I was looking for *mirror equivalence* through a process I called *SRSR* (square–reverse–square root). Upon doodling as far as 100,000, I found two types of integer that give mirror equivalence: when the square was a palindrome (21 in total) and when it was not (132). The integers are listed below.

I also encountered some other interesting results. There were nine integers that, when squared, yielded a palindrome that was not the square of the reverse of the original number (that is, they didn't produce mirror equivalence). And I found six instances where the result neither stemmed from a palindrome nor exhibited mirror equivalence. These integers are also listed below.

Palindromic mirrors

1, 2, 3, 11, 22, 101, 111, 121, 202, 212, 1001, 1111, 2002, 10001, 10101, 10201, 11011, 11111, 11211, 20002, 20102

Non-palindromic mirrors

12, 13, 21, 31, 102, 103, 112, 113, 122, 201, 211, 221, 301, 311, 1002, 1003, 1011, 1012, 1013, 1021, 1022, 1031, 1101, 1102, 1103, 1112, 1113, 1121, 1122, 1201, 1202, 1211, 1212, 1301, 2001, 2011, 2012, 2021, 2022,

2101, 2102, 2111, 2121, 2201, 2202, 2211, 3001, 3011, 3101, 3111, 10002, 10003, 10011, 10012, 10013, 10021, 10022, 10031, 10102, 10103, 10111, 10112, 10113, 10121, 10122, 10202, 10211, 10212, 10221, 11001, 11002, 11003, 11012, 11013, 11021, 11022, 11031, 11101, 11102, 11103, 11112, 11113, 11121, 11122, 11201, 11202, 12001, 12002, 12011, 12012, 12101, 12102, 12111, 12201, 12202, 13001, 13011, 20001, 20011, 20012, 20021, 20022, 20101, 20111, 20112, 20121, 20122, 20201, 20211, 20221, 21001, 21002, 21011, 21021, 21101, 21102, 21111, 21201, 22001, 22002, 22011, 22101, 22102, 22111, 30001, 30011, 30101, 30111, 31001, 31011, 31101, 31111

Palindromic non-mirrors

26, 264, 307, 836, 2285, 2636, 22865, 24846, 30693

Non-palindromic non-mirrors

33, 99, 3168, 6501, 20508, 21468

More fun with number sequences

What is the missing number in the truncated sequence {1, 2, 3, 11, 22, ?}? One might be tempted to answer 33. That's a legitimate answer. But so is 26. For the sequence thus continued can then be interpreted as the set of positive integers whose square is a palindrome. The sequence continues {... 101, 111, 121, 202, 212, 264, 307 ...}, being a blend of what we have called palindromic mirrors and palindromic non-mirrors.

Even without considering the immense flexibility of polynomial sequences—discussed earlier in this book—one is tempted, somewhat timidly, to make the bold conjecture that *all* finite sequences have more than one logical, rule-based continuation.

E: Answers

Cubic polynomials and {1, 4, 9, 16, ?}

We noted earlier that polynomials of degree 2, 4, 5, 6—and probably all the way to ∞—can generate a continuation of the sequence {1, 4, 9, 16 ...}. The question we posed was why cubic polynomials—those of degree 3—are excluded. The answer is quite simple. It is not that cubic polynomials can *never* generate sequences. They obviously can. The cubic polynomial $y = x^3$ clearly generates a sequence. It starts {0, 1, 8, 27 ...}. Rather, given the starting terms of *some* sequences, cubic polynomials cannot generate them. And one such set of starting terms happens to be {1, 4, 9, 16 ...}.

Consider the corresponding subtraction triangle:

$y =$	1		4		9		16
(1)		3		5		7	
(2)			2		2		
(3)				0			

Notice that the first zero starting term is at subtraction level 3. Notice too that the level 3 sequence could be continued as a zero sequence without affecting the starting terms. But level 3 could also be a sequence other than a zero sequence without affecting the starting terms. Let's consider both of these mutually exclusive possibilities

Assume, firstly, that level 3 is a zero sequence. You might think that that makes level 4 also a zero sequence. And if you have a zero sequence at level 4, then a cubic polynomial should be able to generate it. But if you revisit table 2 on page 34, you will see that the equation for creating the first term of the cubic polynomial (6a) would make a equal to 0 (zero being at the start of level 3). Now the first term of a cubic polynomial is ax^3. But if $a = 0$, there is no first term of power 3. The generating polynomial reduces, at most, to a quadratic.[1]

Now if level 3 is not a zero sequence, some number other than zero must appear in it. Let's assume that it is the second number in the sequence (although this is not essential to our argument). Now if you have two non-identical numbers next to one another in a subtraction sequence, clearly their *difference* cannot be zero. Thus subtraction sequence 4 — which must have in it a number that is the difference between a zero and a non-zero number — cannot be a zero sequence. If subtraction sequence 4 cannot be a zero sequence, no polynomial of order 3 — that is, no cubic equation — can generate it. So whether level 3 in the subtraction triangle for {1, 4, 9, 16 ...} is a zero sequence or not, no cubic polynomial can generate it.

It is important to realise that we are stymied here by the actual starting terms of the given sequence. We are not stymied, though, in finding polynomials of greater degree that can generate the given sequence.

From these considerations we can conjecture that if the starting terms of a sequence generate a subtraction sequence of only zeroes at level n, then it will not be possible for a polynomial of degree n to generate the sequence.

---

1. Unless, of course, you want to call a polynomial with a zero coefficient in front of the x^3 term a cubic polynomial. But in that case, every polynomial will have an infinite number of degrees, since a subtraction triangle can be extended downwards indefinitely with an infinite number of consecutive zero sequences, each corresponding to an increasingly more complex polynomial. Thus, say, a quartic polynomial could have 2 as its highest power. And that, at the very least, is a contradiction!

Digital summing

You may have guessed, given the vast number of possible arrangements of the slips of paper, that there must be a short cut to answering this puzzle. And there is! A little experimentation will show that a number is divisible by 3 if the sum of its individual digits is divisible by 3. Consequently, the order in which the slips of paper are drawn out of the bowl is unimportant. For the sum of the digits in the numbers 1 to 20 (which, incidentally, is not the same thing as the sum of the numbers from 1 to 20) is 102. And since this sum is divisible by 3, any number composed—in any order whatsoever—of the 31 digits that make up the numbers from 1 to 20 must itself be divisible by 3. So the probability that the 31-digit number formed is divisible by 3 is 1—a dead certainty!

Patterns before your eyes

Your task was to construct an algorithm that, beginning with the first grid on the left, generated the second and third grids and also the grid labelled A. There are bound to be numerous possibilities, but here is one. If you didn't come up with this particular solution, it only amplifies the suspicion we all should have that intelligence tests are not necessarily what most people think they are.

In the answers below, I call jumping to the top row or to the leftmost column *cycling*. This occurs when a move would force an object outside grid.

Answer A

- The perpendicularity symbol is moved one column to the right.
- If no cycling occurred at the last move, the black block is moved one column to the right, otherwise it remains where it is.

- Each object is moved down one row.
- If such movements would take an object below the grid it is placed on the top row, and if such movements would take an object to the right of the grid it is placed in the first column.
- Each object is rotated 90°clockwise.

Answer B

Your job this time is to construct an algorithm that, beginning with the first grid on the left, generated the second and third grids and also the grid labelled B. Here is one of many possibilities:

- If any step in the previous iteration of this algorithm caused the cycling of either object, move the block one square to the right and the perpendicularity mark one square up; otherwise move both objects one square to the right. If such a movement would take an object to the right of the grid, place it in the first column of the same row; if such a movement would take an object below the grid, place it in the top row of the same column.
- Move both objects down one row. If such a movement would take an object below the grid, place it in the top row of the same column.
- If the objects are in adjacent cells—horizontally, vertically or diagonally—rotate them 90°anticlockwise; otherwise rotate them 90°clockwise.

Answer D

Your job this time is to construct an algorithm that, beginning with the first grid on the left, generated the second and third grids and also the grid labelled D. Here is one of many possibilities, thus proving that any one of the four suggested answers in the multiple-choice question we have been exploring is just as true as any other!

- If the any step in the previous iteration of this algorithm caused the cycling of either object, move both objects one square to the right and one square down (and begin a new iteration of this algorithm, that is, start again from this step); otherwise move both objects one square to the right. If, in either case, such a movement would take an object to the right of the grid, place it in the first column of the same row; if such a movement would take an object below the grid, place it in the top row of the same column.
- Move both objects down one row. If such a movement would take an object below the grid, move it to the top row of the same column.
- Rotate both objects 90°clockwise.

In each case, note that from the fifth iteration onwards the patterns generated by one sweep of an interaction repeat every fourth iteration.

Prime steps

Here is one solution. There may well be many more.

3049

1049

1009

1609

1607

7607

A perfect difference

The two primes are 1999 and 2027, the difference between them being the perfect number 28.

Firstly, note that one of the numbers must start with 2. This is because the numbers, though intermingled, are nonetheless presented in the right order. (That's what the question tells us.) Next consider the various lengths this number could be and the corresponding length of the other number. Since there are 8 digits intermingled, the maximum length that any one number can be is 7 digits, with the other number being one digit. Let's consider this possibility first and assume that it is the seven-digit number that starts with 2. If so, this number can only be 2,199,097 or 2,199,029, with 2 or 7 respectively as the second number (given that 2 and 7 are the only single-digit primes in the original string). Now you can, if you wish, check whether the two seven-digit numbers are primes (which might be a tedious process without a calculator or maths app). Or you can simply note that the only possible difference between them and their corresponding single-digit prime is 2,199,095 and 2,199,022 respectively, and neither is a perfect number (as the sequence of perfect numbers given at the start of the puzzle shows).[2] The only other possibility if one prime is seven digits long is for the single-digit prime to be the 2 at the start of the original string of digits. This would make the seven-digit number 1,990, 297. This too is not a prime, nor is the difference between it and 2 a perfect number. So the larger of the two primes cannot be seven digits long.

There is an easier way to show that a 7–1 distribution of the digits in our original set can never yield a perfect number as the difference. Whatever the single digit, the larger number would have to start with 1 or 2 and thus be in either the one millions or two millions. So too must the difference. And yet, as the sequence of perfect numbers

---

2.   In fact, neither 2,199,097 or 2,199,029 is a prime number.

given at the start of the puzzle shows, there is no perfect number in the one millions or two millions.

Could the larger of the two primes be six digits long? If this number started with 2, the difference between it and the smaller number would necessarily be in the two-hundred thousands and no perfect numbers appears in that range. If the smaller number started with 2, then it can only be 29 (since that is the only two-digit prime that starts with two in the original set of digits). Thus the larger number would start with 1 and be in the 190 thousands, as would the difference between the two numbers. Again there is no perfect number in the one hundred thousands.

Similar reasoning rules out the larger of the two primes from being five digits long. If the number began with 2, the difference between it and the smaller number would be in the twenty thousands, a range once again that is devoid of perfect numbers. If the smaller number begins with 2, it can only be 227 or 229 (since these are the only three-digit primes that start with 2 in the original set of digits). Thus the larger number would start with 1 and be in the 190 thousands, as would the difference between the two numbers. Again there is no perfect number in the hundred thousands.

The only option left is for both primes to have four digits. If so, the only perfect numbers that could be the difference between two 4-digit numbers from the original set are 6, 28 and 496. (This is because the next perfect number after 496 is 8128, and if this was the difference between two four-digit numbers one of which must start with 2, the other four-digit numbers would have to be greater than 9999 — a contradictory and thus impossible situation.)

Notice that the three possible perfect numbers end in 6 or 8. Thus the difference between our two four-digit primes must end in 6 or 8. To continue: the two prime numbers, if they are both four digits long, can only end in 9 or 7. If the larger prime ends in 9, the smaller must end in 3 or 1 if the difference is to end in 6 or 8. But 3 is not in the original set and 1 is too near the start of the set to be the end of a four-digit number. If the smaller prime ends in 9, the larger must end in 5 or 7 if the difference is to end in 6 or 8. Only 7 is in the original set of digits. Put this possibility to one side for a moment and consider the case where the smaller number ends in 7. In this case, the larger number must end in 3 or 5 if the difference is to end in 6 or 8. But neither 3 nor 5 is in our original set. If the larger number ends in 7, the smaller number must end in 1 or 9. We've ruled out 1. Notice that we have eliminated all possibilities except where the larger number ends in 7 and the smaller ends in 9. And if the difference between these two numbers is a perfect number, it cannot be 6 or 496, but 28.

We now have to solve the equation $(abc7) - (def9) = 28$, from which it follows that $c - (f + 1) = 2$ or $(c + 10) - (f + 1) = 2$. Simplified, we have $c - f = 3$ or $(c + 10) - f = 3$. Clearly, selecting from only the digits in the original set, c must be 2 and f 9. We now have $ab27 - de99 = 28$. The digits we have remaining are 2, 1, 9 and 0. Since one of the prime numbers must start with 2—the first digit in the original set—and $abc7$ is larger than $def9$ (established in the previous paragraph), it follows that a is 2 and d is 1. Thus $2b27 - 1e99 = 28$, which is true only if $b = 0$ and $e = 9$. QED.